Superconductors

Conquering Technology's New Frontier

Superconductors

Conquering Technology's New Frontier

Randy Simon
and
Andrew Smith

Plenum Press • New York and London

Library of Congress Cataloging in Publication Data

Simon, Randy.
   Superconductors: conquering technology's new frontier.
   Bibliography: p.
   Includes index.
   1. Superconductors. I. Smith, Andrew, 1953–   . II. Title.
TK7872.S8S55   1988   620.1'12973   88-17950
ISBN 0-306-42959-4

To Paul and Mike
who got us started

Preface

In a laboratory in Leiden in 1911, a Dutch scientist observed the remarkable disappearance of all electrical resistance from a thin capillary of mercury metal sitting in a bath of liquid helium. Seventy-six years later, in a laboratory in Huntsville, Alabama, another scientist saw the same thing happen to a greenish ceramic pellet sitting in a much warmer bath of liquid nitrogen. These two events—the original discovery of superconductivity and the recent discovery of high-temperature superconductivity—are linked by a rich history of scientific and technological accomplishments.

Over the years, superconductivity has flourished as a field of scientific endeavor, leading to the awarding of eight Nobel Prizes in physics. In addition, it has emerged as a technology, contributing to advances in medicine, electronics, astronomy, transportation, and experimental science. But in spite of all these things, superconductivity has remained a little-known phenomenon on the periphery of science.

Now, because of the most recent discoveries, superconductivity has become one of the most celebrated areas in all of science. Its applications promise to reshape many aspects of future society. For the first time ever, superconductivity has become a subject that we should all know about.

We have written this book in order to bring the story of superconductivity to an audience beyond students and practitioners of physics and engineering. In some sense, we have three stories to tell. In the first part of the book, we explore the nature of superconductivity, its history, and our theoretical understanding of the phenomenon. In the second part of the book, we survey a wide variety of practical uses for superconductivity throughout society and examine superconductivity's role in today's economy. In the last part of the book, we discuss the recent breakthroughs in superconductivity and evaluate their impact on the future of technology.

We intend this book for readers who have no prior background in physics, electronics, or other pertinent technical fields. We include no mathematics and presume no familiarity with the principles of modern physical science. We therefore introduce whatever relevant scientific background information we require during the course of each chapter. Although we present modern science without mathematical rigor, we have endeavored to provide an accurate and comprehensive explanation of the subject.

Superconductivity is now a rapidly changing field, with new discoveries being made almost all the time. Although this book includes a wealth of up-to-date information, the very latest news on superconductivity will undoubtedly be found in daily newspapers. This book will help to place the newest results in their proper perspective.

During the process of writing *Superconductors*, we have drawn upon the considerable expertise of a number of colleagues from the superconductivity community. Among these are John Clarke, Robert Fagaly, Sadeg Faris, Eric Forsyth, Ken Grey, Eric Gregory, Robert Hazen, Scott Kreilick, Arnold Silver, Michael Superczynski, Joe Thompson, Harold Weinstock, and Dave Woody. We thank them all for their kind assistance.

A number of colleagues read preliminary drafts of the chapters and provided comments and suggestions. For this important help we thank Paul Chu, Roger Davidheiser, Bill Dozier, Edgar Edelsack, William Gallagher, Robert Hein, William Keller, Vladimir Kresin, Michael Melich, Martin Nisenhoff, and Stuart Wolf. We particularly wish to thank Rachel Cohon and Chris Platt for their efforts in helping to find the best ways to explain difficult concepts.

We wish to express our gratitude to Anne Kottner at the Niels Bohr Library for locating many of the photographs used throughout the book and to Victor Swayne, whose skillful and vivid illustrations greatly enhance the comprehensibility of the book. Finally, we grate-

fully acknowledge the patient but firm guidance of Linda Regan and
Victoria Cherney at Plenum, who kept us honest in our presentation
of a complicated subject.

Randy Simon and Andrew Smith

Redondo Beach, California

Contents

CHAPTER 1

In from the Cold

A silvery glint appears on the horizon, catching our eye. Moments later it passes us with scarcely a sound. We have just observed a sleek passenger train traveling 300 miles per hour while floating several inches above its tracks.

A few miles away, an inconspicuous fenced-in area marks the location of the new power company installation. Buried underground are giant coils of special wire carrying huge electrical currents. The currents flow incessantly, keeping electrical energy stored for when it is needed.

In a nearby hospital, doctors are examining a patient who has suffered a mysterious seizure. An unlikely contraption hovers over the patient's head while the doctors gaze intently at a computer screen. They are closely watching abnormal brain activity taking place near a small tumor.

At an industrial park, some engineers are putting their latest achievement through its paces: the heart of a new computer, faster than any machine ever built and small enough to fit in a shoebox. The microscopic electronic switches within the new computer can turn on and off a trillion times every second.

A fanciful depiction of a distant future? Perhaps. Wishful thinking about undiscovered technologies? Definitely not. All these marvels can be achieved with superconductors and we already know how to do it.

What are superconductors and why haven't we already harvested their rewards?

Superconductors are materials—usually metals—that undergo a

remarkable transformation at extremely low temperatures. Superconductors have a number of unique characteristics but their most notable property and the origin of their name is their ability to conduct electrical current with absolutely no loss of energy. Nothing else in nature can perform this feat.

Superconductors are hard to find but, paradoxically, they are not at all rare. Nearly a quarter of the natural elements are superconductors. There are also hundreds of compounds and alloys with the property. So why are they hard to find? The answer is that superconductivity only happens at temperatures far below anything we are likely to encounter in our lives.

For most people, low temperature means the freezing point of water: 32° Fahrenheit or 0° Celsius. The hardy souls inhabiting the polar climes are even familiar with the one temperature at which the two scales read the same number: $-40°$. The only lower temperature most of us ever come across is that of dry ice at 108 below zero, Fahrenheit. Yet there are still lower temperatures that can be reached. In fact, there is a lowest temperature in nature. According to present-day physics, nothing in the universe can get any colder than 459 Fahrenheit degrees below zero ($-273°$C). Scientists refer to this temperature as "absolute zero" and have devised a temperature scale—the Kelvin scale—for which this lowest temperature is simply zero. On this temperature scale, water freezes at 273 Kelvin and boils at 373 Kelvin.

Until quite recently, the only way to make a material superconduct was to reduce its temperature almost all the way to absolute zero. It may safely be said that reaching such temperatures has never been nor will ever be an easy thing to do.

In 1986 and 1987, pandemonium broke out in the scientific world when a new class of superconductors was discovered. They had the same properties as the old superconductors with one crucial exception: they become superconducting at much higher temperatures. How high? Nearly 100 Kelvin. If you do a little arithmetic, you will realize that this is still plenty cold: about $-280°$F. So why all the fuss?

The answer is primarily economic in nature. Temperatures close to absolute zero are expensive to reach and expensive to maintain. The warmer it gets, the cheaper it gets—that's the rule. It turns out that 77 Kelvin—the temperature of liquid nitrogen—is an affordable and trouble-free temperature in today's technology. If superconductors can operate at such a temperature, then their jobs are waiting. It may sound facetious but in terms of practicality, a 77 K superconductor is truly a high-temperature superconductor.

The discovery of the new superconductors started a mad scramble in the scientific community. A race was on to produce the materials, improve their properties, and exploit them for practical applications. Suddenly superconductivity became a household word, the subject of magazine cover stories and television news reports. The United States Congress and the President called for a national program in superconductivity research. A contagious wave of excitement swept through the physics community. Superconductivity became news but it isn't new at all.

A Dutch physicist named Heike Kamerlingh Onnes lowered the temperature of mercury metal enough so that it became a superconductor back in 1911. He was the first one to do it because only he had the means of reaching such low temperatures. How low? Mercury becomes a superconductor at 4 degrees above absolute zero. The only way to reach such a low temperature is with liquid helium. We know that ordinary steam turns into water if it is cooled below 212°F. In the same way, helium, the gas we use to blow up balloons, becomes a liquid at 4.2 K. It's quite a trick to liquefy helium, but once we've done it, we have a 4-degree fluid. Kamerlingh Onnes became the first person in history to liquefy helium in 1908. Three years later he discovered superconductivity.

Onnes realized almost immediately he was on to something important. He envisioned wire made from superconductors that could be made into huge coils and used as powerful electromagnets requiring no energy to operate. The ensuing decades brought the discovery of new superconductors and new properties of superconductors. Every new discovery presented a new opportunity for useful applications. The only rub was the requirement for temperatures near absolute zero.

So high-temperature superconductivity was an unexpected discovery but an extremely welcome one. Scientists had decades worth of good ideas for using superconductors. They just needed warmer superconductors to make their ideas more practical. It was really no surprise, then, that the 1987 Nobel Prize in physics was awarded to K. Alex Mueller and J. Georg Bednorz of IBM's Zurich laboratory who synthesized and tested the first of the new high-temperature superconductors.

It was also no real surprise that within a few months of the discovery, thousands of scientists around the world were studying the new materials trying to understand how they work and how to improve them. Previously staid, low-key scientific conferences became tumultuous media events with turnaway crowds competing for scarce seats. Universities, government labs, and industrial labs all entered the race for breakthroughs in high-temperature superconductivity.

But for most us, all this uproar about superconductivity is still rather mystifying. We hear things about electrical currents with no loss of energy, we see photos of superconducting magnetically levitated trains, and we read about high-speed superconducting supercomputers. Superconductivity is obviously an important and interesting phenomenon, but just what is it?

To understand anything about *super*conductivity, we have to know something about ordinary conductivity. All ordinary materials resist the flow of electricity to a lesser or greater extent. Electricity or electrical current consists of the movement of electrons through a material. In order to make electrons flow, we need to add energy; we need to push them. Electrons in a metal behave like water in a hose. The water won't flow unless we supply some pressure. The amount of pressure it takes depends on the kind of hose we use.

Similarly, the amount of energy needed to produce a certain electrical current in a material depends on the kind of material we are using. The energy we supply to electrons—the push we give them—is called *voltage*. We get it from a battery or from an outlet in the wall. A 1.5-volt AA battery can supply a certain amount of energy to each electron in a wire. A 6-volt camera battery supplies four times that energy.

The amount of voltage we apply to a conductor determines how much electrical current will flow. The more we push on the electrons, the faster they move. We measure current in units called *amperes* (amps for short). One ampere of current corresponds to more than 10,000,000,000,000,000,000 (usually written as 10^{19}) electrons passing by every second, about the amount of current that flows in an ordinary 100-watt light bulb.

The actual amount of current we get for a given voltage depends on the nature of the material we are using. The more the material resists the flow of electricity, the more we have to push to keep current flowing. We call the measure of this property the resistance of the material. If we apply 1 volt across a material and we get 1 amp of current to flow, then by definition the material has a resistance of

1 *ohm*. If only half as much current flows, then the material has a resistance of 2 ohms, since it resists the flow of electricity more strongly.

Resistances of common materials vary wildly. A chunk of copper—the metal used in most common wire—might have a resistance of only a few millionths of an ohm. A piece of glass, on the other hand, can easily have a resistance in millions of millions of millions (that's quintillions!) of ohms. Almost nothing in nature has as much diversity as the resistance of materials. But all ordinary materials have one thing in common: they have *some* electrical resistance. It takes at least some amount of energy to permit current to flow.

Ordinary materials require voltage to sustain current because they burn up energy in the form of heat. Electrical resistance is a form of friction like the heat generated by tires skidding on pavement. A car stops moving when it loses its energy of motion to heat in its brakes and tires. Similarly, electrons lose their motion in a metal by heating up the metal. We call this loss of energy *Joule heating*.

Superconductors are extraordinary because they alone are immune from the effects of Joule heating. It takes no energy at all to make current flow in a superconductor, and no energy is lost to friction to sustain the current either. Its electrical resistance is precisely zero.

This property is truly sensational, but two questions immediately come to mind: how do we know that the resistance is really zero, and why should the difference between zero ohms in a superconductor and a millionth of an ohm in copper matter to anyone?

The first question almost sounds silly. Zero is zero, after all. But zero really is not an easy number to measure. All instruments have their limits and zero is below the resolution of any measuring device. A human hair would weigh nothing on a bathroom scale, not because it is weightless, but because its weight cannot be detected by the scale. Nevertheless, not even the most careful experiments using the most sensitive instruments have ever found any electrical resistance in a superconductor. These experiments have shown that if there was any resistance at all in a superconductor, it would be at least a trillion times smaller than the resistance of copper. Furthermore, scientists have established a theory that explains superconductivity, and according to the theory, the resistance of a superconductor truly is zero.

So we'll accept that superconductors have no resistance whatsoever. Still, who needs them? Why shouldn't we be satisfied with a few millionths of an ohm in copper? The answer is an issue of energy and scale. Joule heating gets worse and worse for higher resistance materials; double the resistance and we double the heat loss. It gets worse much faster as the amount of current flowing increases. If we double the current in a wire, we quadruple the amount of energy lost to Joule heating.

But copper has such a low resistance. How much energy is lost from a millionth of an ohm? This is where scale comes in. A 1-inch cube of copper has a resistance of a millionth of an ohm. Take the cube and stretch it some to make a block 2 inches long and half an inch wide. What is its resistance now? Four times greater. Why is that? We've done two bad things for the resistance: we've made the conductor longer—so the electrons have more opportunities to lose energy to friction—and we've made it narrower, so their path is more constricted. You can't flow as much water through a narrow pipe as you can through a wide pipe. Electricity obeys the same rules. This produces the problem of scale. Even the resistance of copper causes significant energy losses if we need either very long wires or very narrow wires. The wire that connects a power plant with a distant city is a very long wire indeed. The hair's-width conductors in a complex microcircuit are very narrow wires. Joule heating makes trouble for us with the best ordinary conductors.

Superconductors are a natural solution to the whole problem. They will be most welcome in any application where we need a lot of wire. Ordinary wire used to transmit electricity from power plants to cities wastes 5 to 10 percent of the electrical energy because of resistance. Superconductors would waste nothing at all. Electrical generators made with superconducting wire would be smaller, lighter, and more efficient. When space and weight are at a premium, as on a ship, the benefits could be enormous.

Kamerlingh Onnes' dream of giant superconducting electromagnets has already come true. Current flowing in a coil creates a temporary magnet, called an electromagnet. If the coil is big enough or the current is big enough, such a magnet can be far more powerful than any permanent magnet (like the ones stuck to our refrigerator door). We can build powerful electromagnets with superconducting wire in a fraction of the space needed with ordinary wire. Even coils made from small diameter superconducting wire can handle large currents without burning out from Joule heating. Such superconducting magnets are used today in hospitals around the world for the important new technique of magnetic resonance imaging, in magnet-

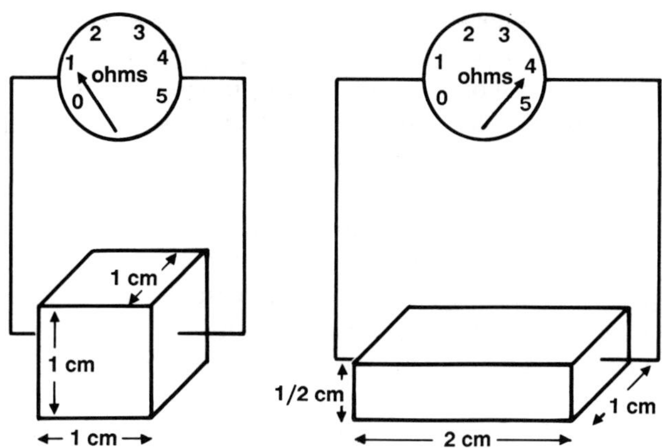

Scaling of resistance.

ic separating plants that can refine metals from ore, and in giant atom smashers and research labs.

The same magnets can levitate whole trains above their tracks and provide part of the magnetic propulsion system that hurdles the train down its track at hundreds of miles per hour. The same zero resistance wire used for magnets can be used in experimental energy storage coils harnessing such powerful magnetic forces that they must be buried in concrete to keep them from collapsing in on themselves.

We have only talked about uses for large quantities of superconducting wire. On the opposite end of the size scale, there are other needs to be filled. The computers of the twenty-first century will have to get really small if they hope to meet the performance demands people will be making upon them. Why small? Because we are rapidly approaching the point at which the slowest thing happening in a computer is the time it takes for the information to travel around the machine. We cannot make information move at a higher velocity because nature limits the speed at which information can travel to the speed of light. The only way to get signals from one end of a computer to the other any sooner is to make the computer smaller. Computers made without superconductors can get only so small before they will literally burn themselves up with resistive heating. Thus, the ultimate computers and other high speed electronics of the future will need superconducting wires.

Superconductors are more than just perfect conductors. They are also perfect diamagnets. *Diamagnetism* refers to a superconductor's ability to repel either pole of an ordinary magnet. If we place a small magnet above a piece of superconductor, the magnet will float above it, suspended in midair. The diamagnetism of superconductors was discovered in 1933 by Walther Meissner; a superconductor's ability to reject magnetism is thus known as the Meissner effect.

The Meissner effect was an important clue in solving the mystery of how materials become superconducting. A great many people spent a great deal of time trying to understand superconductivity. Even after Meissner's work, it still took more than 20 more years and other important discoveries before anyone could really explain it. The physicists who framed the successful theory of superconductivity, John Bardeen, Leon Cooper, and Robert Schrieffer, won a Nobel Prize for their efforts.

The theoretical understanding of superconductivity unleashed another series of discoveries. New properties promised by the theory led to entirely new applications for superconductors. A new kind of electronics has been developed based on such exotic-sounding superconductive circuit elements as Josephson junctions and SQUIDs (Superconducting QUantum Interference Devices). These devices exploit properties of superconductors that we haven't even touched on yet. But what do we gain from superconductive electronics? Circuits made from Josephson junctions and SQUIDs can run faster, use less power, take up less space, and in other ways outperform their conventional electronic counterparts. They have demonstrated world-record performance in high-frequency amplifiers, sensitive radio receivers, detection of minute magnetic signals, and high-precision measurements.

An all-superconducting supercomputer has been a goal for 20 years. A concerted developmental effort is now underway in Japan that may make the superconducting computer a reality in the 1990s. Such a machine will outperform any computer now on the drawing boards.

The same kind of superconducting circuitry is already being used by radio astronomers to detect the faintest signals from the outer reaches of the universe and by medical researchers studying the inner workings of the human brain. Geologists are using SQUID electronics to look for oil deposits deep within the earth's crust. Engineers look forward to advanced satellites filled with complex superconductive circuitry requiring a tiny fraction of the power needed for conventional electronics.

The list of applications of superconductivity is long and varied. Ingenious scientists have had over 75 years to devise ways to exploit the phenomenon. As we have said, many of these uses have become commonplace today. But most remain tucked away in research proposals, unrealized because of the complication of extremely low temperatures.

Superconductors behave like ordinary metals until they reach their "critical temperature," at which their resistance drops precipitously to zero. As we have seen, mercury, the first superconductor discovered, has a critical temperature of 4 K. Other elements were found to have higher critical temperatures, the rather obscure metal niobium taking the prize at 9 K. Over the years some compounds and alloys were found to superconduct at temperatures as high as 23 K ($-418°F$). This is actually "warm" enough to require only liquid hydrogen (which boils at 20 K) for a coolant rather than liquid helium. Unfortunately, liquid hydrogen is not much more attractive than helium. It isn't all that much cheaper, nor easy to handle, and the hydrogen gas that boils off is highly explosive. At least gaseous helium is utterly harmless.

From 1973 to 1986, the 23 K figure remained at the top of the list of superconducting critical temperatures. Many scientists tried to find better superconductors; some even claimed to have succeeded but their results could not be reproduced. Some theorists even tried to prove that there *couldn't* be higher temperature superconductors. The theory of Bardeen, Cooper, and Schrieffer explains how superconductivity works but it doesn't predict what substances should superconduct. Thus the search for higher temperature superconductors was often conducted in darkness. There seemed little cause for optimism.

The work in Zurich in 1986 turned it all around. Mueller and Bednorz synthesized a complicated compound containing four elements and found superconductivity at the lofty temperature of 35 K. This launched a new era in superconductivity. On January 28, 1987, Maw-Kuen Wu and Paul Chu were working with a compound related to the Zurich superconductor in Wu's laboratory at the University of Alabama. The new compound was also a complicated ceramic material, this time made from copper, oxygen, barium, and the little-known rare earth metal yttrium. When Wu attached electrical leads to a greenish pellet of the substance, he observed a transition to the superconducting state using only liquid nitrogen as a coolant. For superconductivity veterans, the Holy Grail had at last been found. The master list of superconductor applications might finally become practical.

Suddenly, the placid field of superconductivity research came alive. The number of people actively engaged in it grew from hundreds to thousands in a matter of months. Numerous newsletters, computer data bases, and new research centers devoted to the subject have sprung up and are flourishing. New industries promise to be created, and old ones to be rejuvenated, all contingent upon the new superconductors making the transition from the laboratory to the outside world. The initial worldwide reaction was one of almost unbridled enthusiasm, a gold rush in science.

As the shock waves wear off from these remarkable discoveries, we can start to realistically assess the consequences of high-temperature superconductivity upon technology and, in turn, upon society. What is unquestionably now a revolution in the world of physics may well turn out to be revolutionary in the world as a whole. For this to happen, tremendous technological challenges must be met, because the new materials are complicated and difficult to work with. What these challenges are, how to meet them, who will meet them, and what the future is likely to hold will occupy the latter part of this book.

But first, to understand the challenges of the new superconductive technology, and to appreciate the contributions of the old technology, we need to learn about superconductivity in general. The whole spectrum of superconducting phenomena is exploited in one way or another for superconducting technology. Understanding these phenomena is the only way to see how superconductivity can play a role in so many applications areas and to see what role it is likely to play in our future lives.

As we've seen already, superconductivity, like any technical field, abounds with its own peculiar technical terms and buzzwords. Josephson junctions, SQUIDs, critical temperatures, and diamagnetism have all been mentioned in the course of just a few pages. We will carefully define these new terms when we need them in our exploration. Also, we have provided a glossary of terms at the end that may be consulted when too many new concepts seem to pile up at one time.

The next few chapters address the science of superconductivity, a branch of the broader discipline of solid-state physics, the study of the properties and behavior of solid matter. Superconductivity usually comes up as a topic near the tail end of a basic course in solid-state physics; in this book we will look at superconductivity without the benefit of such an introduction. To do so, we'll simply pick up the necessary scientific basics along the way. At the same time, we'll get a more human perspective of the field as we trace the historical development of superconductivity.

With this background, we will then proceed to survey the whole sweep of superconducting technology: from levitated trains to magnetoencephalographs, from satellites to laboratory measurement standards, and from supercomputers to supercolliders. We will see where superconductivity already is at work and where it is ready to step in. And finally, we will examine the headline-making events of the recent past and look ahead to the nearly unlimited promise of a superconducting future.

CHAPTER 2

No Resistance—No Magnetic Field

Superconductors intrigue us because they do many wondrous things, but they especially intrigue us because they do one thing perfectly: conduct electricity. Few things in nature exhibit true perfection but superconductors carry current with absolutely no resistance. Although, as we have seen, no measurement technique can distinguish between zero and some very small number, we have strong reason to believe that the resistance in a superconductor really is zero. This simple departure from ordinary electrical behavior is what makes superconductors truly super. But what constitutes ordinary electrical behavior?

Electricity flows through wires. We don't think much about it, it just happens. We flip a switch and a light comes on. Yet it was less than 100 years ago that Sir Joseph Thomson discovered what electricity is made of: negatively charged electrons. We now say simply that "electrons flow through wires." But why this happens, where the electrons come from, and why, for example, the electrons only flow through the wire—and not through the plastic insulation surrounding the wire—are tougher questions.

Even casual observers could see that some substances are better at carrying electrical current than others. Metals like copper, silver, and gold are clearly the best. Scientists understood that current would not flow by itself; an "electromotive force"—a battery or voltage source—was needed. The simplest understanding of electricity draws an analogy to water in a pipe. You can fill a pipe with water but in order to make the water flow you have to apply pressure. In the same way, a copper wire contains plenty of electrons but requires the pressure of a voltage in order to make electricity flow.

But what does voltage really mean? Voltage measures the potential energy we give to each electron in a wire. Potential energy is energy that is stored up and ready to do work. There are familiar examples. When we compress a spring, we store up potential energy that is released when we let the spring go. If we raise a weight up to some height above the ground, we give it gravitational potential energy that is released when we let go of the weight. In both cases, the potential energy can be converted to kinetic energy, which is energy of motion. Similarly, the potential energy we give to electrons by applying a voltage can be converted into kinetic energy by letting the electrons travel through an electrical circuit.

Ordinary Resistance

The electrical properties of matter are a part of solid-state physics: the study of solid matter, its interactions, and behavior. Early in the twentieth century, solid-state physicists first studied the mechanism by which electricity flows in materials. With their early theories, they could actually predict such things as which substances should conduct electricity like a metal and which substances should insulate like glass. They could calculate the effects of applying a voltage across a metal wire. But there was a flaw in their picture of the motion of electrons in a metal. They predicted that once the electrons started flowing, there would be no reason for the current to stop. Sir Isaac Newton had worked out his laws of motion in the seventeenth century. According to Newton, an object in motion in the absence of an external force would continue moving forever. Theoretically, electrons in a metal seemed to be subject to no force that could slow them down. Unimpeded, electrical current should flow through a metal indefinitely. Batteries shouldn't be necessary to keep it flowing.

A theory that predicts something we know to be false has a major problem. In this case, the solution is not too mysterious. Anyone who has run electrical current through a wire knows it: the wire gets hot; friction dissipates the current. It actually took a while to figure out where the friction comes from, but its existence was indisputable. In fact, we exploit that very friction at every opportunity. A light bulb, an electric heater, or a toaster, for that matter, simply contain specialized pieces of wire that heat up from the friction of the electrical current passing through them.

As noted, the heat generated by friction in wires is called Joule heat and the friction itself is called *resistance*. We have also noted that the electrical resistance of different materials varies widely; some have a lot, some have very little, and superconductors have none at all. In fact, solid-state physicists find it convenient to group materials according to their resistance to electricity and thus we have insulators, semiconductors, metals, and superconductors. Whether or not we are aware of it, the first three groups play important roles in our lives and have for a long time (although semiconductors have only come into their own since the late 1950's). Superconductors, on the other hand, have mostly remained laboratory curiosities for decades.

What distinguishes these groups of materials from one another? Metals have large numbers of electrons that are not tightly bound to particular atoms but rather are shared by many atoms and are free to roam about in response to external influences. Solid-state physicists call wandering electrons *itinerant electrons*. These mobile charge carriers make metals good conductors of electricity.

The electrons in insulators, on the other hand, are not free to move about in this way but rather are bound to particular sites in the solid. Great amounts of energy are required to yank electrons away from their parent atoms. Thus, insulators strongly resist the flow of electricity.

In between metals and insulators are semiconductors that have a small number of electrons that can conduct electricity. Silicon and germanium are the most familiar semiconducting elements. Semiconductors have proven to be very useful because the addition of minute amounts of extra elements can drastically change their ability to carry electrical current. By precisely controlling the presence of such impurities in miniature regions of semiconductor materials, engineers have developed devices like transistors and integrated circuits. Such semiconductor devices are the building blocks of our modern electronic technology.

Nevertheless, the story of superconductivity is a story about metals. As we have said, the sticking point in the early theory of electronic motion in metals was the origin of resistance. The itinerant electrons in a metal, once sent on their merry way by the force applied by a battery, seemed to have no reason to slow down or stop.

One could just as well take away the battery and expect the current to keep going. And yet the electrons clearly lose energy to the material in the form of Joule heat. So how do the electrons lose their energy?

The answer requires a little understanding of the structure of solids. Most solids are comprised of atoms arranged in a very orderly fashion consisting of neat rows and columns forming many layers. Such an arrangement is known as a *crystalline lattice*. There are a number of types of such lattices in solids and an entire science—crystallography—is devoted to their study. Metals, semiconductors, and insulators are all made up of lattices. Nearly perfect specimens of such crystalline lattices are quite beautiful to behold; diamonds and sapphires are familiar examples of such single crystals.

The first successful theory for the electrical properties of metals was actually proposed in 1900 by Paul Drude and Hendrik Lorentz, shortly after the discovery of the electron. According to the theory, a truly perfect crystal of a metallic solid would in fact conduct electricity without any resistance. Unfortunately, even the edges of a crystal constitute a kind of imperfection with respect to the geometric regularity of the lattice; the repeated pattern of atoms comes to an end at the edges, which, like the walls of a handball court that reflect a ball, present obstacles to traveling electrons. So to test this part of the theory, we would need an infinite crystal. Of course we can't really have an infinite crystal, but even if we did, it would probably still be imperfect in other ways. All real crystals have at least some impurities—atoms of the wrong kind—or irregularities—atoms in the wrong position—so that there will always be some obstacles to passing electrons. But even if we somehow managed to produce a perfect, infinite crystal, there still remains another mysterious source of resistance. To see what this is, we need to understand something about how solids are held together.

The atoms in a solid are arranged in their orderly lattice by virtue of their electric charges with which they exert forces upon one another. These forces are not like solid rods linking the atoms; think of them more like springs, subject to expansion and compression. What can cause these springs to expand or compress? The main culprit is temperature.

When we measure the temperature of an object we are observing the amount of random motion of its atomic constituents. The higher the temperature the more the atoms are jiggling around inside the material. If this motion grows too violent in a solid, the atoms in the

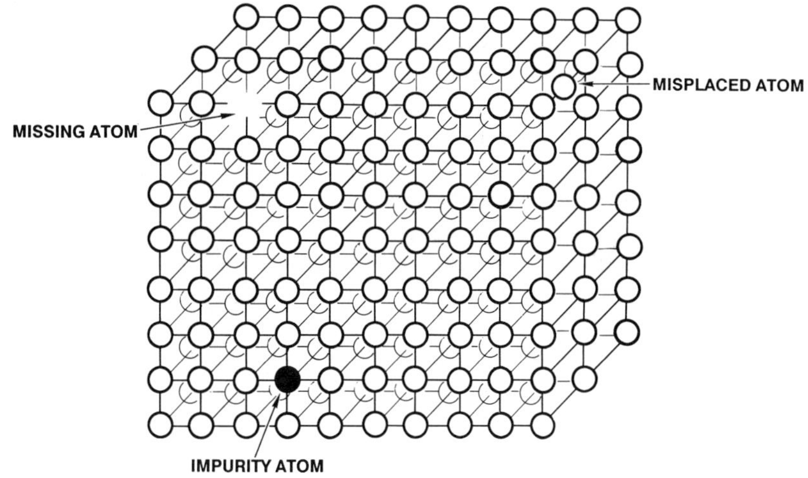

Crystalline lattice.

lattice can no longer keep their orderly arrangement and the whole structure comes apart. The solid melts.

Under less extreme circumstances the spring forces that hold the atoms in place in a solid allow them to move about from their regular or equilibrium positions. The analogy with springs is not just for convenience. The behavior of atoms in a solid is really quite like balls connected by springs. In particular, if we take a system of balls and springs and set it into motion, it behaves in a very specific way. The balls oscillate back and forth only in certain patterns and at certain frequencies. Two balls connected by a spring will oscillate at only one frequency. As we add more balls and springs, we find more ways that the system is willing to move. Such characteristic patterns of motion are called normal modes.

By the 1920s, physicists had developed a new theory for the behavior of the submicroscopic world of atoms and electrons called *quantum mechanics*. The phenomena of the older "classical" physics soon were matched by quantum mechanical counterparts. The quantum mechanical equivalent of the normal modes of a solid are entities called *phonons*. As we will see, phonons play a major role both in ordinary conductivity and in superconductivity. But whether we speak of phonons or of lattice vibrations, the important physical principle is that at any finite temperature, the atoms in a solid move around a little bit. Temperature supplies thermal energy to the atoms, which excites the natural vibrations of the lattice. Moving atoms are no longer in their proper locations, so we no longer have a perfect crystal lattice.

The theory of solids predicts no resistance in a metal with perfect crystal order but real solids always have imperfections. Missing atoms, atoms of the wrong kind, edges of the crystal, and places where the rows and columns somehow get disarrayed are all examples. But even without these imperfections, the motion of the atoms alone leads to resistance. Scientists had eliminated the mystery surrounding the origins of resistance. In the language of quantum mechanics, we say that an "electron–phonon interaction" has taken place, but this is just a fancy way of saying that moving atoms interfere with electrons.

We have already said that temperature excites the motion of the atoms in a solid. So, according to this view, as the temperature of a metal is lowered, the thermal vibrations of the atoms should decrease and the resistance should fall. What about the motion of the electrons themselves? If the electrons are really free to move about in the metal, low temperatures shouldn't matter. They don't get their energy from the influence of temperature to begin with. Thus, this picture of electronic motion in a metal tells us that if a specimen were perfect, its resistance would approach zero as its temperature approached zero. In actual metals, there would always be some "residual resistance" because of impurities and other imperfections. Kamerlingh Onnes, the man who discovered superconductivity, was simply trying to test this theory on the purest metal sample that he could get his hands on.

Superconductors Have No Resistance

The discovery of superconductivity in 1911 defied all predictions of the theory of metals. When Onnes cooled mercury down to the temperature of liquid helium, the resistance didn't smoothly decrease toward its residual value as expected from the theory; it abruptly dropped to zero. No matter how sensitive a measurement Onnes tried to make, he found no resistance left in the mercury. And, worse yet for the theory, adding impurities to the mercury didn't change these results at all.

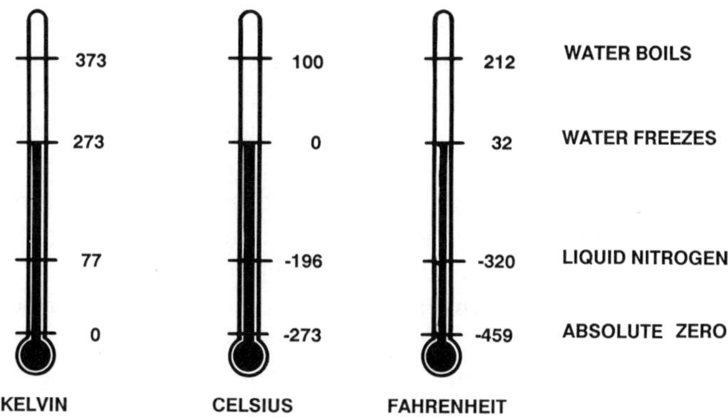

Comparing temperature scales.

Scientists soon discovered that this amazing property was not unique to mercury. Tin and lead behaved in the same way, although the dramatic disappearance of the resistance occurred at a different temperature for each. A number of other metals showed this super-conducting behavior as well, but just as interesting, others did not. In fact, the best ordinary conductors—copper, silver, and gold—have never superconducted at any temperature that experimenters have reached in their laboratories.

As we have seen, the temperature scale that physicists use is known as the absolute or Kelvin scale where zero denotes the lowest temperature possible in nature (actually a temperature that can only be approached but not reached). In this scale the degrees are the same size as Celsius degrees, so that water boils 100 Kelvin degrees above its freezing point. To convert from Kelvin to Celsius, you simply subtract 273 degrees from the Kelvin value. Thus absolute zero (0 K) is $-273°C$ (or $-459°F$, if you prefer). This may seem rather confusing and not worth the bother, but if you work with superconductors, it is far easier to say that lead superconducts at 7 K rather than $-266°C$ (or $-446°F$). Therefore, from here on, we will often quote temperatures in Kelvin; keep in mind that small numbers in Kelvin mean *really* low temperatures.

Superconductivity could not be discovered until Kamerlingh Onnes first liquefied helium in 1908. Helium becomes a liquid at 4.2 K, a temperature extraordinarily close to absolute zero. Using liquid helium as a refrigerant, scientists could make measurements at unprecedented low temperatures. The first superconductor, mercury, loses its resistance right at the liquid helium temperature. So in those early years, a "high-temperature superconductor" meant lead, with its somewhat greater 7 K superconducting transition temperature. Of course, all such temperatures are fantastically lower than anything we have experienced ourselves; that fearsome 40-below night in Minnesota tops out at a balmy 233 on the Kelvin scale.

We used the term *transition temperature* when we referred to the superconductivity in lead. That is because when a metal becomes a superconductor, we say that a phase transition has occurred, in this case from the so-called normal state to the superconducting state. We are familiar with a number of phase transitions, such as water freezing to form ice or boiling into steam. In each case, a material changes its state of matter or phase when a certain temperature is reached. A phase transition in a given material takes place at a characteristic temperature (for example: 0°C for freezing water and 100°C for boiling it).

Thus, the phase transition that changes an ordinary metal into a superconductor occurs at a characteristic temperature that is different for each particular superconductor. This transition temperature is also called the *critical temperature* and, in fact, is often written for short as "T_c." Therefore, we would say that lead has a T_c of 7 K, meaning that when we cool a piece of lead down to that temperature, its resistance will abruptly drop to zero.

The superconducting transition typically takes place over a fairly narrow temperature range. The resistance starts to precipitously fall at some temperature and by the time the temperature drops a fraction of a degree lower, no more resistance can be measured. The difference in temperature between the start of the drop and where zero occurs is called the *width of the transition*. It varies with the purity of the metal we are measuring, but is generally only a very small fraction of a degree. In any case, when the transition is over, the resistance really is zero. How can we be sure?

We can never actually *prove* that the resistance is zero. We could imagine building a new extraordinarily sensitive instrument to detect incredibly small resistances and trying it out on our superconductor. If we read zero with our instrument, we could still only say with certainty that the resistance is less than the smallest number the instrument can register. As we said before, this is the same problem as trying to weigh a hair on a bathroom scale. On the other hand, no

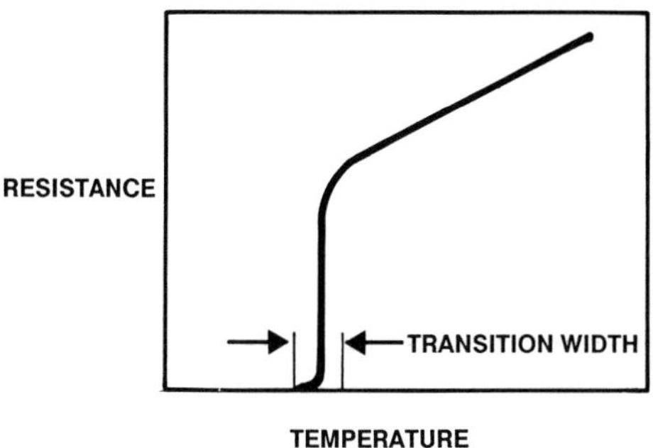

Superconducting resistance transition.

experiment has ever been done that has found any resistance in a superconductor.*

The best test for zero resistance is to make a closed ring of superconductor and induce a current to flow in the ring. We can do this by moving a magnet into the loop—the same principle by which an electrical generator operates—since changing magnetic fields generate electrical currents. However, nature behaves symmetrically and flowing electrical currents in turn generate magnetic fields. This fundamental law of electromagnetism allows us to build electromagnets by passing currents through coils of wire. Thus, the current in a superconducting ring turns the ring into a magnet. The influence exerted by a magnet—called a magnetic field—is something that we can measure quite sensitively. If the magnetic field produced by the current in our superconducting ring doesn't change, then the current in the loop hasn't changed. If the current stays the same, then there is no resistance in the loop. Proving that something hasn't changed is no different from proving that something is zero; in the second instance we still have to show that the change is zero. The advantage in the magnetic experiment comes from our superior ability to measure tiny magnetic signals. Thanks to highly sensitive electronics (superconductive electronics, no less), scientists can detect incredibly small changes in the magnetic field from a superconducting ring. With such measurements, they can set a far smaller upper limit on the resistance of a superconductor than can be deduced from any direct resistance measurement. In fact, superconducting ring measurements show

*As we will see, this is only strictly true for direct current, not for alternating current.

that if any resistance remained in a superconductor, it would have to be less than a billionth of a billionth the resistance of the best ordinary conductor. Occam's Razor clearly threatens to cut away this last remnant; in other words, we should accept the simplest hypothesis that explains the data: the resistance is zero.

A more intuitively appealing way to test for zero resistance is to get the current flowing in the loop, wait a long time, and see what happens. If there was any resistance in the loop, the current would die away. This is one of the most boring experiments in the world to do; people have let it go on for years without anything happening to the current. One such famous experiment only ended when the person in charge of replenishing the liquid helium in the flask containing the superconducting ring forgot to do his job one night. The helium boiled away, the ring warmed up, and the experiment was over.

Such persistent currents might seem like perpetual motion and, in some sense, they are. No one can build a real perpetual-motion machine because we can't take more energy out of a machine than we put in to run it. However, no laws of nature are being violated by superconductors. We don't extract any energy from the current in the loop, therefore we are not "getting something for nothing." It turns out that nature is filled with these kind of ersatz perpetual-motion machines. The electron in a hydrogen atom is an example. It goes around and around the nucleus forever; it is a persistent current. But it is legal because the orbiting electron does not do any work or deliver energy anywhere during its travels.

The zero resistance property of superconductors strictly holds only for direct electrical current. Electricity comes in two forms: direct current (dc) and alternating current (ac). Direct current flows in one direction only, while alternating current periodically reverses its direction. The current that flows in household wiring is alternating current, reversing its direction 120 times every second. Every two reversals result in the current heading in its original direction, so we describe such current as being 60-cycle-per-second current.

Superconductors do not have zero resistance to the flow of alternating current. In fact, the higher the frequency of the ac electricity, the higher the resistance in the superconductor. From a practical point of view, however, the resistance in a superconductor at low frequencies (like the 60-cycle-per-second currents in our homes) turns out to be so low as to be insignificant, even as compared to good conductors like copper. But technically speaking, only direct currents can flow with absolutely no resistance at all.

The existence of the zero-resistance state of superconductivity prompted two important lines of inquiry. Why does it happen, and does it have any useful applications? The search for answers to such questions often only leads to even more questions. Such was the case for early investigators of superconductivity.

Superconductivity and Magnetism

The Scottish physicist and mathematician James Clerk Maxwell developed a theory that was a crowning achievement of 19th-century science: the classical theory of electromagnetism. Culminating in four interrelated equations, Maxwell's theory ties together the phenomena of electricity, magnetism, and light in a profound and elegant manner. Moreover, Maxwell's electromagnetic theory had remarkable predictions to make about superconductors decades before they were discovered.

Before we can understand what Maxwell's theory predicts about the relationship between superconductors and magnetic fields, we need to understand the concept of a magnetic field. If you hold a magnet under a piece of glass covered with iron filings, the filings line up in patterns of loops extending from one pole of the magnet to the other. These so-called lines of force show us the presence and location of a magnetic field surrounding the magnet. All magnets generate such lines of force. The lines of force that emanate from the earth's magnetic north pole and make up our planet's magnetic field can be traced out with the needle of a compass. The concept of a magnetic field allows modern science to avoid the philosophically disturbing idea of action-at-distance. Objects simply don't influence one another over distances with nothing connecting them. We know that a magnet can attract a piece of iron that is placed some distance away from it. Only a mystical view of the phenomenon would assert that the magnet simply acts across the intervening space. Instead, we can more reasonably say that a magnetic force permeates the space in the vicinity of the magnet; the space contains a magnetic field.

Most substances are more-or-less transparent to the presence of magnetic field lines. For example, field lines pass freely through the glass upon which we sprinkle iron filings. Some materials become magnetized themselves when strong fields pass through them; permanent magnets are made in this way. But Maxwell's electromagnetic theory says that something unusual will happen when we place a superconductor—or more accurately, a perfect conductor—in a magnetic field.

According to one of Maxwell's fundamental laws of electromagnetism, a perfect conductor will not permit any change in magnetic field to occur within it. What does that imply? It says that if we bring a magnet near a superconductor, the field lines can't get in, for if they *were* to enter the superconductor, the magnetic field obviously would not be the same as it was before. Such a change is forbidden by the theory. So what do we expect will happen?

Our glass and iron filings example can help answer the question. If we bring a piece of ordinary copper near the magnet in our first experiment, nothing happens to the filings. They form the same elaborate patterns as before. If we instead bring another magnet next to the first one with like poles adjacent, something odd happens. The field lines traced out by the filings are pushed out around the second magnet; they cannot enter it. In mathematical terms, the two fields add together to form new field lines of a different shape; this is known as superposition. In physical terms, what happens to the magnets is that they repel each other.

This is exactly what the superconductor will do in proximity with a magnet. It rejects the magnetic field that is applied to it and thereby repels the magnet. But with the superconductor, it doesn't matter which pole of the magnet we use. With either end the result is the same. Something that repels any pole of a magnet is called a diamagnet (where the prefix *dia* means opposite).

So, the electromagnetic theory of the nineteenth century was able to predict an interesting property of superconductors: if you place one in a magnetic field, it excludes the field. However, in 1933, Walther Meissner and Robert Ochsenfeld made the second crucial discovery in the history of superconductivity.

In the early 1930s, there was a great deal of interest in the nature of the superconducting state. Were superconductors simply perfect conductors that could be described by Maxwell's electrodynamics, or were they something different? Was superconductivity really a phase transition like boiling or melting? For reasons that will become apparent later, we can answer the question by seeing what happens if we do the magnetic field experiment backwards. In other words, instead of producing a magnetic field in the presence of a superconductor, we try inducing superconductivity in the presence of a magnetic field. What was supposed to happen? According to Maxwell's equations,

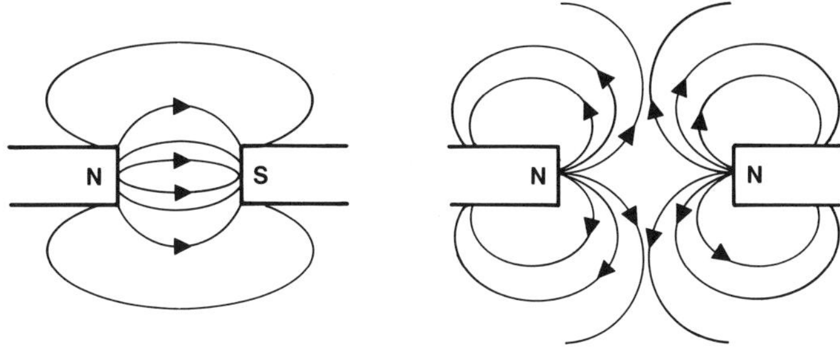

Magnetic field lines near attracting (left) and repelling (right) magnets.

the field shouldn't change in the perfect conductor. Why not? If a material becomes a perfect conductor in the presence of a magnetic field, Maxwell's law tells us that the field must remain constant in the conductor. Magnetic fields cannot change in a perfect conductor. So, if we cool down a block of lead next to a magnet, the magnetic field inside the lead should be trapped in when the lead becomes a superconductor. Take away the magnet and the field should still be there, unchanged and unchanging, for if it went away, Maxwell's law would be violated. But this means that superconductors are not necessarily diamagnets; if we lock in a magnetic field, we will have magnetized the superconductor. We should then be able to attract another magnet instead of repelling it.

Surprisingly, Meissner's experiment showed that this was not what happens at all. The end result of performing the magnetic experiment backwards is the same as performing it the first way. In either case, we get a superconductor with no magnetic field inside. This was a remarkable result. In one case the superconductor screens out the magnetic field that we apply; in the other case it expels the field that was there in the first place. A superconductor is not only a perfect conductor (which allows it to screen out an applied field), it is also a perfect diamagnet—it always expels magnetic fields. Later we learn how this behavior helped scientists to understand the nature of superconductivity. This magnetic response of superconductors is now known (perhaps unfairly to Ochsenfeld) as the Meissner effect.

Scientists did not invent the concept of diamagnetism for superconductivity. The phenomenon occurs quite commonly in most atoms and molecules. Diamagnetism also plays a central role in the operating principles of motors and generators. A basic law of electromagnetism—Lenz's Law, named after the nineteenth-century German physicist Heinrich Lenz—describes what happens when we try to change the magnetic field that passes through a loop of wire. If we start with some magnetic field and then change it, a voltage appears across the wire causing a current to flow in the loop. With this principle, we can build electric generators. However, as we also know, the induced current that flows generates a magnetic field. According to Lenz, the new field always tries to resist the change that was made in the original field. This circular sequence of events constitutes the essence of diamagnetism. Succinctly, diamagnets fight off magnetic fields by generating currents that produce opposing fields. Add extra magnetic field in one direction and the material tries to create that amount of field in the opposite direction. Ordinary copper acts as a diamagnet, but its electrical resistance causes the induced currents to quickly die off. Thus ordinary conductors try to avoid changing magnetic fields, but they rapidly fail as their diamagnetic currents disappear. Such currents could only persist if they flowed without loss of energy. The same diamagnetic response happens in atoms and molecules, in circuits, and, it turns out, in the Meissner effect. But only superconductors can win the battle against a changing magnetic field.

These two effects—perfect conductivity and perfect diamagnetism—are the hallmarks of superconductivity. No one could hope to understand superconductivity until both effects had been discovered. Thus, it is not surprising that the real development of a theory of superconductivity only began after the discovery of the Meissner effect.

The Strength of Superconductivity

In order to understand superconductivity we need to realize that it is not an all-or-nothing phenomenon. When superconductivity occurs at T_c, the zero-resistance state is very fragile—it is easily destroyed—and strengthens only as the temperature is lowered. But how can superconductivity be destroyed? Strange as it may seem, we can do it by applying electrical currents or magnetic fields. Superconductors can only withstand so much current or so much magnetic field before losing their unique characteristics. We thus introduce two new properties of superconductors: critical currents and critical magnetic fields.

We know that ordinary wires heat up when electrical currents flow through them. If we put too much current in a wire, the wire will burn up. To prevent this, we use thicker wire. Why does that help? What matters to the current-carrying ability of a wire turns out to be what is called the *current density*—the amount of current that flows through a given area of wire. Make the area greater and more total current can flow through a wire made of the same material. So we use thicker wire to spread out the current.

But what happens in a superconductor? The current doesn't cause any heat because the resistance is zero. So it seems that we could put all the current we want into a superconductor without burning it out. Unfortunately, it doesn't quite work that way. Once again, there is a maximum current density for a given kind of wire. To the advantage of superconductivity, this limit is far greater than any ordinary wire could handle without vaporizing. Yet there is a maximum current that even a superconductor can carry. The mechanism for destruction, however, is different than for ordinary wire. The wire doesn't gradually heat up. What happens instead is that the superconductivity abruptly vanishes when we turn the current up too high. We call the current at which this happens the *critical current*. Notice that if the large current that caused the wire's superconducting properties to vanish is not turned off, Joule heating sets in with a vengeance. The wire's days are truly numbered.

The second way to destroy superconductivity is with magnetic fields. Apply a strong enough field and, once again, superconductivity vanishes. The field at which this happens is known, not surprisingly, as the *critical field*. But wait a minute. We just finished explaining the Meissner effect and we learned that superconductors expel magnetic fields. Now we are saying that magnetic fields destroy superconductivity. What is going on? The answer is that if you apply a large enough magnetic field, you can forget about the Meissner effect. The field will catastrophically enter the superconductor and destroy the superconductivity in the process. When we discuss a few more aspects of superconductivity, we will be in a position to understand why this happens.

It turns out that the critical field constitutes a more fundamental property than the critical current. We say this because, at least in the elemental superconductors, the critical current actually results from the critical field. As we have said, when current flows in a wire, a magnetic field is generated. Apply more current and produce a greater field. When enough current flows in a superconductor to generate a magnetic field equal in magnitude to the critical field, then superconductivity is destroyed. Thus the critical current really comes from the critical field.

Both of these new properties are important for understanding how superconductivity works. They are also important for demonstrating the way the strength of superconductivity changes with temperature.

What do we mean by "the strength of superconductivity"? We refer to the resilience of the superconducting state against being destroyed. If we try destroying superconductivity with current, for example, we find that the critical current changes with temperature. It takes a different amount of current to destroy superconductivity at different temperatures. This amount is greatest at temperatures far below T_c and is quite small near T_c. In fact, at exactly T_c, the critical current for any superconductor becomes infinitesimal. And, moreover, at exactly T_c, the critical field for any superconductor also becomes infinitesimal.

We will eventually see that all the important properties we can measure in superconductors are very dependent upon temperature. When we lower a superconductor's temperature further below its transition point, it seems in a sense to get more superconducting as it gets colder. Eventually this trend tapers off. By the time we get down to about half of the critical temperature, the critical current, the critical field, and a number of other properties nearly reach their maximum values. In other words, the superconductivity is as strong as it can get.

Near the transition temperature, on the other hand, superconductivity is as fragile as can be. At T_c the resistance of a superconductor vanishes but the superconductivity is not at all resilient. We can destroy it with minute currents or magnetic fields. We need to lower the temperature of a superconductor in order to produce "full-strength" superconductivity. As we will see later, this kind of behavior was a major clue in solving the mystery of the origin of superconductivity.

When a material becomes superconducting, nothing dramatic seems to happen to it physically. By the time materials reach superconducting transition temperatures, whatever other effects that low temperature might produce—such as solidification, increased rigidity, and contraction of the lattice—have already been observed. Right at the superconducting transition, the physical structure of the material stays the same, the density stays the same, the color stays the same, and so on. But the resistance goes away. As we know, electrical currents are carried by electrons. So, the key to superconductivity must be that something happens to the electrons. They become superconducting electrons. And just what are superconducting electrons? It took many years to answer that one. In fact, John Bardeen, Leon Cooper, and Robert Schreiffer won a Nobel Prize for providing the answer. We will examine their work later on.

For now, the important thing to know about superconducting electrons is that they provide a satisfying explanation for the way superconductors get stronger as they get colder. The first real theory of how superconductivity works was called the two-fluid model. This model says that below T_c a superconductor contains a mixture of two kinds of electrons: normal and superconducting. One simply postulates that there are no superconducting electrons above T_c, very few near T_c, and lots of them at very low temperatures. The more superconducting electrons there are, the stronger the superconductivity. How many can there eventually be? Certainly no more than the total number of electrons we started with in the first place. Superconductivity cannot create new electrons. With the two-fluid model, we can explain how superconductivity starts out weak and levels out to a maximum strength with decreasing temperature. So something must happen in a superconductor that turns normal electrons into superconducting electrons, and it is these superconducting electrons that are responsible for all the tricks that superconductors can do.

Incidentally, this distinction between superconducting electrons and normal electrons provides an explanation for why superconductors are not perfect conductors of ac current. Both kinds of electrons are ordinarily present in a superconductor. When a dc current starts to flow, all the electrons try to carry it. But current will literally take the path of least resistance and the superconducting electrons soon take over the job of carrying all the current. If currents have to repeatedly change direction, however, the situation changes.

All electrons—whether superconducting or normal—have the property of inertia; they resist changes in motion. Thus, when ac current repeatedly forces electrons to reverse their course, they do so with reluctance. At the crucial moments when electrons must come to a halt and turn around, superconducting electrons and normal electrons are back on an equal footing. Both types again try to carry the current. At these moments, the normal electrons dissipate energy in the form of Joule heat. The substance develops some resistance. Clearly, the more often the electrons need to reverse their direction, the more resistance will be developed. Thus at higher and higher frequencies, superconductors lose more of their advantages in carrying electricity. At lower frequencies, on the other hand, the normal electrons carry current so seldom that the resistance developed remains negligible.

The two-fluid picture of superconductivity actually goes a long way towards explaining the phenomenon. The brothers Fritz and Heinz London developed a similar theory about superconducting electrons in the 1930s that embodied within it both the zero-resistance property and the Meissner effect. Theoretical solid state physicists of the time considered this a major breakthrough.

The London theory originated yet another key concept associated with superconductors: the magnetic penetration depth. The penetration depth relates to the diamagnetism in superconductors. As we have seen, diamagnetism comes from induced currents that flow in the presence of magnetic fields. According to the London theory, the diamagnetism in a superconductor comes from currents formed by superconducting electrons on the surface of the superconductor. If we apply a magnetic field to a superconductor, these currents flow in order to generate a magnetic field that can cancel out the applied field. Ordinary conductors respond in the same way, but superconductors can sustain diamagnetic currents indefinitely.

The idea of a magnetic penetration depth comes about in the following way. A region exists on the surface of the superconductor where these diamagnetic currents actually flow. Here the magnetic field cannot be zero because the currents generate a field. However, this region does not have to be very deep. In fact, for most superconductors, the magnetic screening currents are limited to a region only a few hundred atomic layers deep. This depth is called the *magnetic penetration depth*.

An ordinary oven mitt provides a simple picture of a penetration depth. When we pick up a hot pan from the oven, our hands stay cool, thanks to our oven mitts. Yet the surface of the mitts must be very hot. What happens is that the heat only gets so far into the material, and then stops. The distance into the mitt that the heat penetrates might be called the penetration depth for the heat. In the London theory, superconductors have a penetration depth for magnetic fields. The theory not only accounts for the existence of this kind of behavior, it also predicts the actual magnitude of the penetration depth. Physicists call this kind of theory a phenomenological theory. That means you can predict what will happen; you just have no idea *why* it happens. Both the two-fluid model and the London theory are phenomenological theories of superconductivity. Both enjoyed their measure of success in explaining what was known about superconductors in the 1930s. What came next, however, demanded more sophisticated theories.

The kinds of superconducting phenomena we've discussed so far are classical phenomena in the sense that the physics developed before the twentieth century has no trouble describing them, although a complete explanation might be lacking. However, there are a number of entirely different aspects of superconductivity that cannot be adequately described by classical physics. Such phenomena are at the heart of many of the modern applications of superconductivity and are crucial for developing a microscopic theory of superconductivity. These fall under the heading of the so-called quantum effects in superconductors. We take them up next.

CHAPTER 3

Superconductivity and Quantum Mechanics

Up to this point we have been talking about relatively garden-variety phenomena: resistance, magnetic fields, and low temperatures. We are now ready to delve into some of the exotica of superconductivity. For this, we will need to know a little about quantum mechanics.

The modern theory describing the behavior of the atomic and sub-atomic world—quantum mechanics—has the disadvantages of having little contact with ordinary phenomena and dealing with a number of rather abstract concepts. Therefore, intuitions we derive from everyday experience can't help us much in developing an under-standing of the quantum effects in superconductors. Nevertheless, some concrete images can help us form a basic picture of the quantum mechanical principles underlying the behavior of superconductors.

Particles, Waves, and Quanta

We start with the central topic in quantum mechanics: the parti-cle-wave duality. We have a basic understanding of the concepts of particles and waves. Particles are little chunks of matter. We look at a particle and can identify what it is, where it is, and perhaps, which way and how fast it is going. Particles such as protons and electrons are the building blocks of atoms.

Waves are conceptually a bit more difficult to explain. But we have all seen them. Ocean waves are the most obvious example. If we sit at one spot in the water, we bob up and down as a wave passes. If we travel on the wave, as on a surfboard, we move across the water at a fixed height. The wave is an undulation of the water that also propagates through it. It has crests and troughs and a speed of travel.

Waves are common phenomena. Light, radio and TV transmissions, and X rays are waves. In fact, they are all electromagnetic waves, oscillating electric and magnetic fields that travel through space at the speed of light. They have amplitudes—a measure of the difference between the crest and trough—and frequencies—the number of crests going by per second.

Particles and waves seem to be two distinct concepts. What do they have to do with one another? Until quantum mechanics came along, very little, as far as anyone knew. But the work of a French prince, Louis de Broglie, done in the 1920s, showed that there are some cases in which particles act like waves and some cases in which waves act like particles. The result depends on what kind of experiment we do, much as our impressions of an object vary depending on which of our senses we use to examine it. This ambiguity between particle and wave is fundamental to the very nature of atoms and subatomic particles. The capacity of particles to act like waves and vice versa plays a major role in the theory of quantum mechanics. It is called the particle–wave duality and is essential to our understanding of many phenomena in the microscopic world.

The theory of quantum mechanics allows us to understand the behavior of physical systems on the atomic and subatomic level. On this submicroscopic level Newton's laws of motion, which predict the behavior of particles, and Maxwell's laws of electromagnetics, which explain the wave motion of light, have gradually proven to be incomplete and incapable of correctly predicting what happens. The behavior of subatomic particles such as electrons instead has to be described in terms of so-called *wavefunctions*, which are mathematical constructions that tell us about the probability for finding an electron at a certain place rather than telling us the electron's exact location. According to this "wave mechanics," an electron actually has some probability of being in any number of places at a given time, which is a rather troubling concept. When we talk about bigger objects, on the other hand, the theory tells us how to combine the wavefunctions for all the atoms in such a way that we finally come to the conclusion that the object is indeed in one particular place. Quantum mechanics provides a probabilistic description of the world rather than a deterministic description. Clearly, this is not a very intuitive theory.

Why should we bother with such a philosophically unsettling construction for the workings of the microscopic world? The answer is that this counterintuitive description works and the older classical mechanics does not. However, for most of us, the behavior of individual subatomic particles is of little concern. And when we get to the behavior of macroscopic objects like cars and baseballs, quantum mechanics (fortunately) ends up with the same results as classical mechanics. This is known as the *correspondence principle* and saves us from having to renounce Newton's Laws of Motion after they have worked just fine for 300 years. Newton's laws do an excellent job of predicting the motion of baseballs, cars, and even planets.

Explaining superconductivity is another matter entirely. There are many aspects of the phenomenon that are manifestations of quantum mechanics on a macroscopic scale. We need quantum mechanics to explain the workings of even macroscopic chunks of superconductor. And quantum mechanics plays an equally important role in the rest of the modern theory of solids.

Besides the peculiar commingling of the concepts of particle and wave, quantum mechanics contains a number of other unique features. The word *quantum* itself applies to the aspect of the theory that asserts that a number of physical properties can only occur in certain discrete units. The physical properties we are familiar with don't seem to work like this at all. Someone could be exactly 6 feet tall or could also be just over 6 feet. Height can change by infinitesimal increments; it is a continuous property. The same example could be made for a person's weight. However, if we focus down at the atomic level, even such things as height and weight are not continuous. Since we know that we are made up of atoms, we realize that our height and weight cannot change by less than one atom's worth (or for sticklers, the size of a subatomic particle). Of course this is an absurdly minute change, but the principle holds. There is a smallest unit; there is a quantum. Perhaps monetary transactions provide the most familiar example of quantization. Prices are measured in units of dollars and cents; we can't change a price by less than a penny.

In science, electrical charge is the most familiar quantized property. Charge always comes in units called "e." Electrons have one negative quantum of charge, $-e$, and protons have one positive quantum, $+e$.

Various other particles have +e, −e, +2e, −2e, or even no charge at all (although the controversial building blocks of particle physics called *quarks* are said to have fractional charges). But no matter what object's charge we decide to measure, we always get some number of e's; charge is quantized.

Another aspect of quantum mechanics divides elementary particles into two categories: bosons and fermions. The two are distinguished by another quantized property called *spin*. Spin relates to a particle's angular momentum. We are familiar with ordinary momentum; objects in motion carry momentum by virtue of their mass and their velocity. Similarly, rotating objects are said to carry angular momentum by virtue of how their mass is arranged and how fast they are spinning. Although we can't really tell whether subatomic particles spin like tops, we can infer from certain measurements whether they have angular momentum. For some unknown reason, most elementary particles in fact do. But even if no one knows why particles have the property of spin, scientists can measure it. The proper units for spin are rather peculiar, but usually we ignore the units and just talk about spin as being either whole-number (or integer) spin or half-number (or half-integer) spin. Thus a particle might have a spin of 1, 2, or −1, or perhaps a spin of $1/2$, $−1/2$, or $3/2$. Exactly what this means is unimportant; a spin of 1 is simply twice as big as a spin of $1/2$. The key point is that physicists call particles with integer spins *bosons* and particles with half-integer spins *fermions*. They make this distinction because, according to quantum mechanics, groups of fermions have to play by a different set of rules than groups of bosons.

Electrons are fermions; each one has a spin of $1/2$. All fermions obey a rule called the Pauli Exclusion Principle, which means that if you get a group of them together in a crystalline solid, no two of them can have both the same energy and the same spin. Another way to say the same thing is that no two fermions can occupy the same quantum state, where a quantum state is specified by listing all the quantized properties of the fermion (charge, spin, energy, etc.). Such a rule resembles the rule that no two people can have the same Social Security number. Interestingly, bosons obey no such rule. Photons, the quanta of light, are the most familiar bosons (they have zero spin). There can be an unlimited number of them in the same quantum state. Thus we can build powerful lasers that produce intense beams of light of a single frequency; all the photons can be in the same quantum state. This fundamental distinction in the properties of fermions and bosons dictates the behavior of systems containing large numbers of particles.

Energy Levels and the Gap

What are the consequences of the Pauli Exclusion Principle? We have already spoken of the electrons in a metal as being itinerant electrons; they are free to move about within the metal. But according to the Pauli Exclusion Principle, they are restricted in the sense that no two electrons can occupy the same quantum state.

To get a feel for what this is all about, we imagine the following model for constructing a metal. We start with the metal stripped of all its conduction electrons, the ones free to wander. We now proceed to add the electrons back to the metal one at a time. We add the first electron. It will have zero kinetic energy. That means it isn't moving. We add the second electron and it also has zero energy but it has the opposite spin from the first (it would be $-\frac{1}{2}$ instead of $+\frac{1}{2}$ but physicists usually talk about "spin down" and "spin up" in these cases). The third and fourth electrons cannot have zero energy because that would violate the Pauli Exclusion Principle. Thus they will have a small amount of kinetic energy. We continue to add more electrons making sure that each pair has a higher energy than the previous one. Eventually, we fill up the metal with all its electrons. The last two electrons must have lots of energy of motion in order to be in a unique quantum state. The amount of energy that the most energetic pair of electrons has gets the special name *Fermi energy* in honor of the famed Italian physicist, Enrico Fermi. If any more electrons were to be added to the metal, they would have to have more than this amount of energy.

All of this sounds pretty abstract but it can be understood in a very ordinary way. An easy way to visualize this process is to think of the metal as a tall skinny container and the electrons as marbles we pour into the container. Each layer of marbles will sit higher in the container when we pour it in. The energy levels of electrons in a metal are analogous to the height of the marbles except that we also can speak of the energy levels independently of the electrons. The levels can exist without the electrons in the same sense that parking spaces can exist without cars. Such vacant electron states indicate where new electrons (or marbles) would go. If we think of the elec-

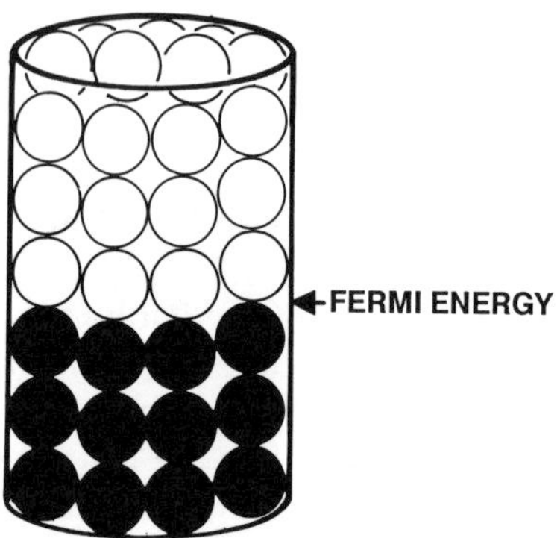

Marbles in a container help us visualize the energy levels in a metal. Black marbles represent filled levels; clear marbles represent vacant levels.

trons and energy levels in this way, then fermions behave in a sensible manner.

Pouring marbles into a container provides a very good analogy when we are trying to explain what is called the *free-electron model*, the simplest way to describe electron behavior. However, real solids are not accurately described by the free-electron model. The energy levels in a solid comprise what is called the *energy spectrum* of the material. In a real material, the states are distributed in a way that is peculiar to that material. What we mean is that the number of states near a certain energy level might be different from the number near another level. This is like having wide parts and narrow parts in our container full of marbles. States would be scarce in the narrow part of the container and dense in the wide part. In fact, this distribution of levels is called the *density of states* in the material.

We end up with the equivalent of an oddly shaped container in real solids because the presence of the crystal lattice causes the simple energy levels in the solid to become much more complicated. The wavelike nature of the electrons when they are confined within a lattice allows them to interfere with one another in much the same way as light waves interfere with one another in an oil slick causing colors to appear in the black surface. In the case of the electrons, we don't get colorful interference patterns, we get energy bands. Instead of a continuous succession of adjacent energy levels, there are groupings of levels called bands.

For most materials, these bands are not continuous. The inter-

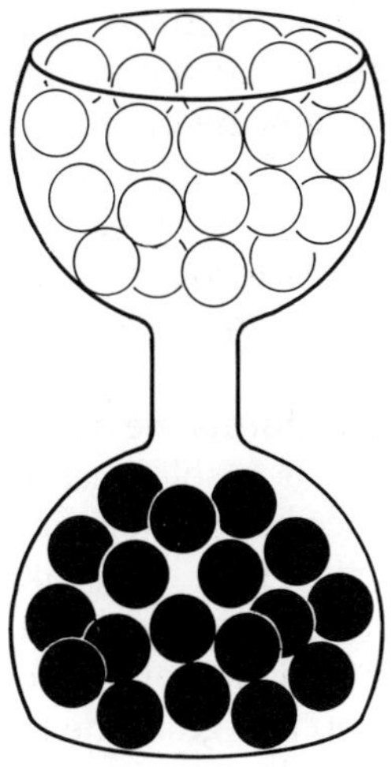

Energy gap between conduction and valence bands.

ference effects of the lattice causes the bands to split away from one another. In that case, when we reach the highest energy level in a certain band, the next available level might be at a much higher energy with no levels in between. We call the last energy band to be filled with electrons the *valence band*. The band above it—full of vacant levels—is called the *conduction band*. We refer to the energy difference between the top of the valence band and the bottom of the conduction band as the energy band gap, or more simply, the *energy gap*.

The energy gap is an interesting and completely quantum-mechanical concept. It represents a forbidden region in energy—sort of a no-parking zone—for electrons. Energy gaps were first detected in semiconductor crystals by observing the way they absorb infrared radiation that is shined on them. Infrared radiation is simply light whose frequency is less than the smallest we can see, which is a shade of red. The infrared spans a broad range of frequencies below visible light. Below a certain frequency in the infrared, light cannot be absorbed at all by a semiconductor crystal. The energy either bounces off the crystal or goes right through it. But when the radiation reaches a certain frequency, absorption suddenly takes place. What happens is that the radiation carries a certain amount of energy corresponding to its frequency. If the radiation carries an amount of energy smaller than the energy gap, then the electrons in the semiconductor cannot absorb that energy because that would require them to occupy a forbidden energy level. When the energy supplied by the radiation becomes greater than or equal to the energy gap, then the electrons can occupy empty levels in the conduction band and the energy can be absorbed. Thus we can directly measure the magnitude of the energy gap by observing the frequency of radiation necessary to permit absorption.

This energy-band model forms the basis for our understanding of insulators and conductors. If a solid has just enough electrons to completely fill up its valence band, then we have an insulator. Why? Because electrical conduction in a solid takes place when electrons move from occupied level to empty level. If all the levels are full, then the electrons can't move and no conduction takes place. But why can't the electrons at the top just go to the next higher level? Because that level sits at much too high an energy in an insulator. Too large a gap separates it from the valence band.

In metals, the valence band is only partially full and there are plenty of empty states to occupy. Electrons can move from full state to empty state and conduction occurs easily in a metal. But the most interesting case is that of semiconductors, the materials that are moderately good at carrying current. Semiconductors are like insulators in that they have filled valence bands. In fact, at absolute zero, they *are* insulators. However, the energy gap in a semiconductor is small enough that just the motion caused by temperature can give the most energetic electrons enough additional energy to jump from the valence band to the conduction band. Temperature plays the same role as the infrared radiation did in providing the necessary energy to bridge the energy gap. So at finite temperatures, semiconductors can conduct some current and do so better and better as they are warmed up. Room temperature provides enough additional thermal energy to excite a good number of electrons into the conduction band of a typical semiconductor. The same thing would work in a real insulator, except that the temperature necessary to bridge the gap energy turns out to be about 10,000 degrees!

All of this band theory is interesting in its own right if we are mostly interested in understanding conductors and insulators, but what about superconductors?

Scientists learned that superconductors have an energy gap. The energy gap in superconductors was discovered in the same way as

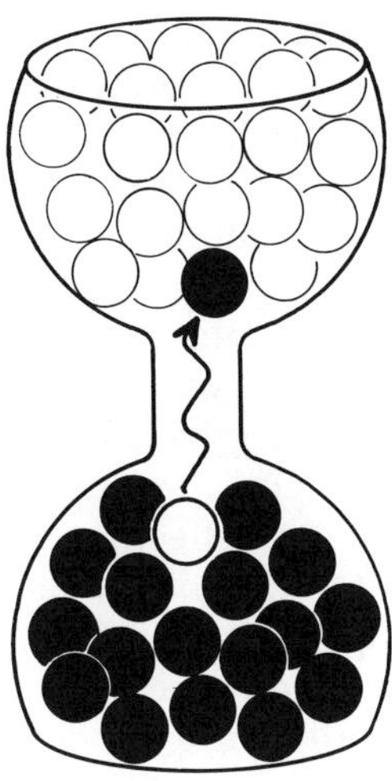

Adding energy to overcome an energy gap.

the gap in semiconductors. When light is shined on a superconductor, it too has an "absorption edge"—a minimum frequency at which radiation is absorbed. However, the frequency at which the absorption occurs in a superconductor is about a thousand times lower than that of a semiconductor. Put another way, the temperature necessary to excite electrons across the superconducting gap is only a few degrees, about the same as a superconducting critical temperature.

The energy gap in superconductors differs from the gap in semiconductors in a very fundamental way. In semiconductors, the gap prevents the flow of electrical current. Energy must be added to take electrons from the valence band into the conduction band before current can flow. In a superconductor, on the other hand, current flows despite the presence of a gap. The energy gap has no effect upon the behavior of the special electrons that carry current in a superconductor. These electrons can be distinguished from normal electrons, as we shall soon see. But superconductors contain normal electrons as well, and it is these electrons that are affected by the gap.

There are other experiments apart from absorption experiments that demonstrate the existence of a superconducting energy gap. One of these experiments consists of measuring the specific heat of a superconductor. To raise the temperature of a substance, we must add energy to it—usually in the form of heat. The amount of energy required to elevate a material's temperature depends on the material. The specific heat measures the amount of energy we have to add to a material in order to raise its temperature by a given amount. It takes more heat to raise the temperature of a high-specific-heat material than it does to raise the temperature of a low-specific-heat material.

Scientists have understood the mechanisms controlling the specific heat in ordinary metals for quite some time. The heat energy we add to a metal can go into the crystal lattice or into the electrons in the metal. Each contribution affects the way a metal absorbs heat at a given temperature, but the total effect can easily be predicted. Superconductors, on the other hand, exhibit unusual specific-heat behavior right at their critical temperature. At that temperature, the specific heat becomes abruptly higher than it was before the onset of superconductivity. In other words, an extra amount of heat is suddenly required to raise the temperature of the superconductor. We can understand this behavior in terms of the existence of an energy gap because once again there are forbidden energy levels that cannot be occupied by the electrons. We can't give the electrons an amount of energy that would place them in a forbidden level. Thus, to raise the temperature of the metal, we now have to overcome the energy gap.

There are, in fact, a number of experiments that suggest the presence of an energy gap in superconductors. The obvious question is why does such a gap exist? Nobody could really explain it until the Bardeen-Cooper-Schreiffer (usually called BCS) theory of superconductivity was developed in 1957. We defer our discussion of this theory until later, where we discuss the origins of the energy gap. But apart from explaining the gap, we should also bring up the most convincing demonstration of its existence: the superconducting tunnel junction.

The Tunnel Junction

The tunnel junction is well worth talking about in detail not just because it is the best way to observe the energy gap in superconductors but because it forms the basis of the entire technology of superconductive electronics. A tunnel junction is physically a very simple thing. Picture a sandwich in which the bread is made of metal and the filling is some kind of insulator. The tricky part is that the insulator

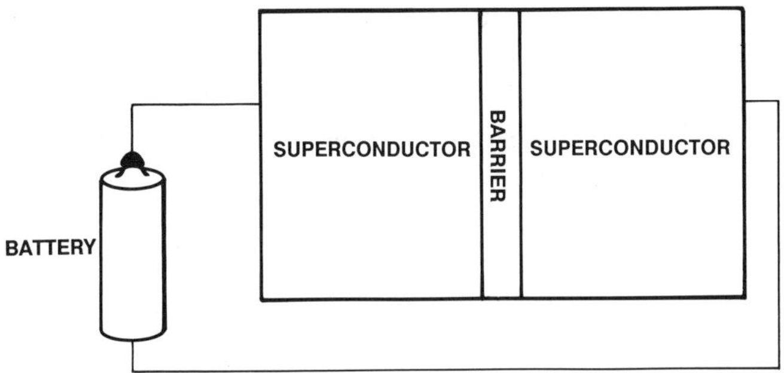

A tunnel junction formed from two superconductors.

has to be unbelievably thin: about 20 angstroms—the thickness of 10 atoms. How one goes about making such a sandwich is an interesting story, but for now, we'll just accept that it can be done.

The thing to do with a tunnel junction once we've got one is to try to pass current through it. We hook up a battery across the junction—from one metal side to the other metal side—and watch what happens. According to classical electrical theory, nothing should happen because we have an insulator between the two pieces of metal, and we know that current can't flow in an insulator. The battery should just force electrons through the metal where they can only pile up, unable to pass through the insulator. Two pieces of metal separated by an insulator that store up electrical charge in this way constitutes a common electrical device called a *capacitor*. Electronic circuits are filled with capacitors, which are workhorses for the temporary storage of electrical charge. They are interesting devices, but have little to do with superconductivity. Nevertheless, according to classical electrical theory, a tunnel junction is a capacitor.

If that was the whole story, we wouldn't be at all interested in tunnel junctions. When we actually try the experiment, however, something unexpected happens: current flows right through the insulator. Once again, in keeping with the theme of this chapter, we can only understand this behavior with the help of quantum mechanics. The wavelike nature of an erstwhile particle allows the electron to do something surprising. The electron can't really go through the insulator, but according to quantum mechanics, it has a finite probability of existing at the other side because the insulator is so very thin. Let's look at this another way to better understand it.

A more physical description of this process reveals that when the electron wave gets to the boundary between the metal and the insulator, it can no longer keep traveling because it has reached a forbidden zone. So the wave dies out—its amplitude gets smaller and smaller. However, it takes a finite distance for the wave to die out completely. A central concept in quantum mechanics—the Heisenberg Uncertainty Principle—has to do with events in nature not happening too abruptly in space or in time. The wave must occupy some of both space and time before it completely dies out. This is where quantum mechanics gives the electron a break. If the amplitude of the electron wave hasn't reached zero when the wave gets to the other side of the forbidden zone, the wave can resume its propagation and the electron goes on its merry way.

Without the help of quantum mechanics, an ordinary insulator won't pass current because it has a large band gap that acts as a barrier. The electron needs too much energy to overcome this barrier. But when it gets across a thin insulator by quantum mechanical sleight of hand, it has gone *through* the barrier rather than over it. For this reason, we say that the electron has tunneled through the barrier.

This tunneling phenomenon is quite extraordinary. It simply cannot happen according to classical physics. And nobody had ever seen it happen until the 1960s. The reason for the long delay was a practical one: the insulator really has to be exceedingly thin and no one knew how to make one. If the insulator is much more than 10 or 20 atomic layers thick, the electron wave dies out and no tunneling takes place. When we discuss practical superconductors, we will see what kind of clever experiment is required to produce the right structure to make a real tunnel junction.

We discuss three types of tunnel junctions: those made of normal metals, ordinary superconducting tunnel junctions, and Josephson superconducting tunnel junctions.

Tunnel junctions may be extraordinary in the way in which they conduct electricity but there is nothing unusual about the current that flows through a junction made of normal metals. Normal metals conduct electricity according to Ohm's Law, which states that the current that flows through a conductor is proportional to the voltage applied across it. This behavior is quite sensible; if we double the voltage, the current doubles. Something that behaves in this way is said to have a linear response. But for any material the ratio of the voltage and the

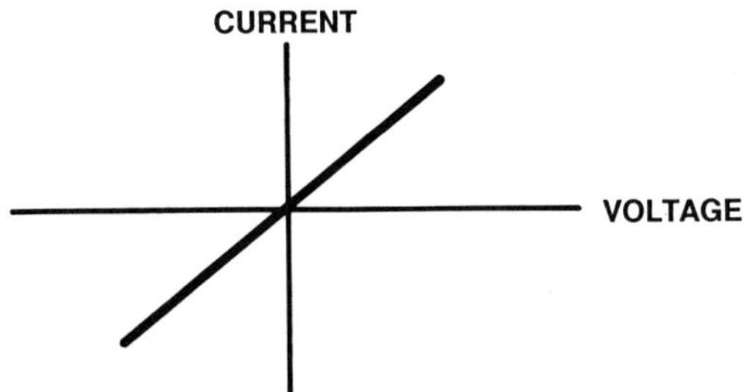

A linear current–voltage characteristic.

current defines the resistance. Thus a material with twice the resistance of another requires twice the voltage to produce the same current.

Electrical engineers draw graphs of the current and voltage behavior of various electronic devices that reveal current–voltage characteristics. An electrical engineer can glance at one of these pictures and immediately tell what kind of device it was drawn for. The current–voltage characteristic of a resistor we might buy at an electronics parts store (typically just a little package containing a piece of carbon or a thin metal film having a certain resistance) appears as a straight line; the device has a linear response. This is simply a pictorial representation of Ohm's Law: the current is proportional to the voltage. If all current–voltage plots looked like this, nobody would ever draw them. Most useful electronic devices have current–voltage characteristics that look nothing like straight lines. Such devices are therefore known as nonlinear devices. When we double the voltage across a transistor, for example, we don't necessarily get twice the current.

Tunnel junctions made from normal metals have straight-line current–voltage characteristics. They obey Ohm's Law as long as the currents are reasonably small. It seems surprising to find that a device that carries current in such a peculiar fashion acts like an ordinary conductor. When we see the reason for this, we will recognize the special advantages of superconducting tunnel junctions.

Tunnel junctions are a direct way to observe the density of electronic states in a material. To see why this should be, we return to our marbles analogy. A tunnel junction can be visualized as two containers full of marbles sitting side-by-side. They are filled to the same level; they have the same fermi energy. We remember that the fermi

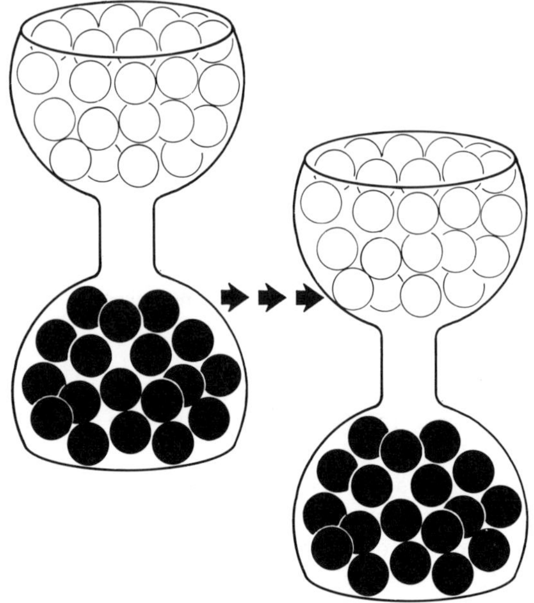

A representation of tunneling at a voltage exceeding the energy gap.

energy corresponds to the top level of marbles and the density of states corresponds to the width of the container. The wider the container is, the more marbles we can place at a certain height; this corresponds to a high density of states. Now, how does tunneling happen?

For the tunneling process to actually take place, there have to be electrons to do the tunneling and there has to be some place for them to go. Electrons that tunnel have to come from filled energy levels and go to empty energy levels. Electrons gain or lose no energy when they tunnel; they have the same energy on either side of the insulator. Because of this, in the absence of a voltage, no tunneling can take place. In terms of our analogy, we can't take a marble from one container and put it at the same height in the other container because there already is a marble in that position.

Therefore, in order to make tunneling happen, we need to change the energy levels in the junction. So, we hook up a battery to the junction, producing a voltage across it. We have given additional energy to the electrons to force them to flow. In our analogy, hooking up a battery is like raising up one container higher than the other—all its marbles sit at a higher position than before. Now a marble at the top of the higher container can go straight across to an empty space in the second container. In the tunnel junction, the electrons in the filled states on one side can now tunnel into the empty states on the other side.

This is where the density of states enters the picture. In normal metals, the density of electron states changes only very slowly as a function of energy. Imagine that our container no longer has vertical walls but instead gently curves as we move higher. Over a small distance, we might not notice any change in the width of the container. In the same way, over a small energy interval we don't notice the effect of the changing density of states in a metal. The number of full and empty states at each successive level is roughly the same. As a result, when we apply a voltage, the number of full states and empty states participating in the tunneling stays about the same. We transfer electrons from levels with similar numbers of occupants. Hence the amount of additional current that flows is proportional to the voltage. If the current is proportional to the voltage, the tunnel junction obeys Ohm's Law.

For superconductors, on the other hand, we have a very different situation. The density of states has a gap in it right at the fermi energy. This means that there is an energy interval over which there are no states at all. We might imagine this as an impassible constriction in our container of marbles. It also turns out that when we get past the gap, there are suddenly a large number of states available. This is known as a peak in the density of states. It happens because the total number of electrons in the metal doesn't change just because we rearrange them. If no electrons are allowed in the gap, then they must pile up at its edge. We would need quite an unusual container to simulate this behavior.

How does this situation affect our tunnel junction? What happens is that when we apply a small voltage to the junction, we don't get any current. Why not? Because of the energy gap. If the energy supplied by the battery is smaller than the energy gap, then electrons cannot tunnel across because of the forbidden zone on the other side. The current–voltage characteristic then looks very different from that of a resistor. We would draw a horizontal line indicating that no current flows for the applied voltage.

When the voltage is high enough, however, we can supply

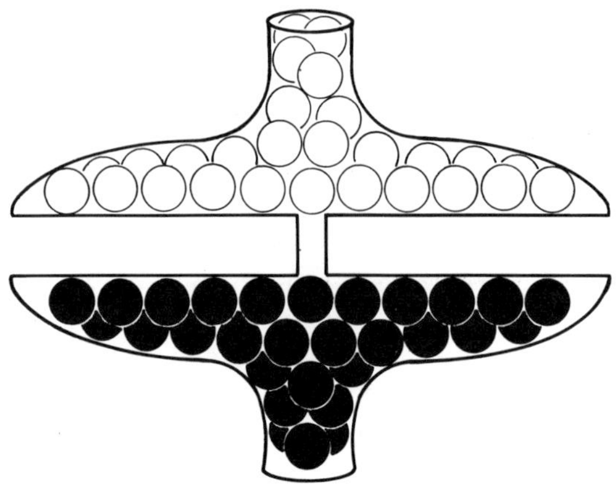

A fanciful container representing the density of states in a superconductor.

enough energy to the electrons to exceed the energy gap. Now there are empty states available to receive the tunneling electrons. In fact, loads of empty states become available because of the peak in the density of states. Actually, there are also loads of occupied states to provide tunneling electrons since filled states pile up below the gap and empty states pile up above it. Therefore we observe that at a certain voltage, we get a large surge of tunneling current. A small increase in voltage produces a big change in current. This is not a linear response. With still more voltage, we get past the peak in the density of states and the current becomes proportional to the voltage once again.

Putting it all together, we find that a tunnel junction made from superconductors has a current–voltage characteristic that looks nothing like a straight line and thus constitutes a nonlinear device. The active elements in electronic circuits—for example, semiconductor devices like transistors, integrated circuits, and diodes—are all nonlinear devices. Such devices are essential for the development of any electronic technology.

We might note that a superconducting tunnel junction, although made of superconductors, does not behave like a superconductor—it does not permit the flow of current without any voltage. The reason for this is that its current-carrying properties are determined by the insulating barrier that forms the junction, not by the superconduc-

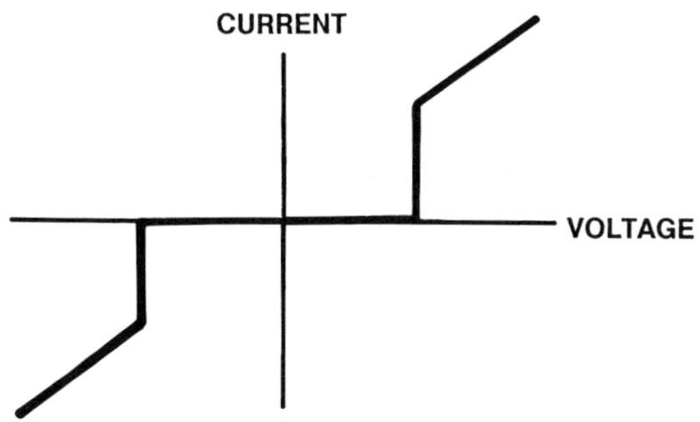

Highly nonlinear relationship between current and voltage in a
superconducting tunnel junction.

tors. Momentarily, however, we will see that there exists another
kind of tunnel junction that *can* behave like a superconductor.

There are two important consequences of the behavior of superconducting tunnel junctions. One is that tunnel junctions give us a
way of looking directly at the density of states in a superconductor
because the current–voltage characteristic shows us how many electrons occupy each energy level. Thus we can observe the energy gap
by measuring the voltage at which a sudden onset of current occurs.
This was and remains to this day the best and most unambiguous
way to observe the superconducting energy gap, a key element in our
modern understanding of superconductivity.

The second consequence of the behavior of superconducting tunnel junctions is more practical than theoretical. We have discovered a
useful new electronic device. Nonlinear devices can be exploited to
perform electronic tasks like rectification—controlling the direction in
which current can flow—and amplification—making small signals
larger. Such functions are the building blocks of more complicated
electronic circuitry. As a result, almost all of superconductive electronics is based upon tunnel junctions, primarily a special type called
Josephson junctions.

Josephson junctions were theoretically predicted before anyone
tried to make one. Superconducting tunnel junctions were first made
in 1960 by Ivar Giaever, a scientist at General Electric's research laboratory. We have already talked about so-called two-fluid models of
superconductors in which electrons are classified as either normal
electrons or superconducting electrons. Soon after the discovery of

superconducting tunnel junctions it became clear that it was the normal electrons rather than the superconducting electrons that tunnel in a junction. Brian Josephson, a Cambridge graduate student, calculated that it ought to be possible for superconducting electrons to tunnel as well. He was soon proven correct and ever since, tunnel junctions that allow superconducting electrons to tunnel have been known as Josephson junctions. Superconducting electrons have a slightly more difficult time in tunneling through an insulating barrier, so Josephson junctions differ from ordinary superconducting junctions in that they have slightly thinner barriers. Apart from this minor difference, ordinary superconducting junctions and Josephson junctions are the same thing; however, superconducting electrons can only flow through Josephson junctions.

So far we have purposely avoided discussing just what superconducting electrons are. When we get to the Bardeen-Cooper-Schrieffer theory of superconductivity, we will see that in fact they are pairs of electrons that are bound together in a special way. For the moment, what we already know about superconducting electrons is sufficient for discussing Josephson junctions. Superconducting electrons are electrons that carry supercurrent—current with no resistance. Thus, if superconducting electrons can tunnel, they should tunnel with no resistance. But no resistance means that no voltage is needed to make current flow.

This is exactly what happens in a Josephson tunnel junction. A current flows with no voltage applied. How can we understand this in terms of our familiar container full of marbles? The answer is that we can't. The superconducting electrons cannot be compared to the marbles in the container. They are a different sort of entity. In fact, for our purposes, superconducting electrons can be thought of as bosons. As we have said, bosons play by a different set of rules. They can all occupy the same quantum state; they can have the same energy and spin just like the light coming out of a laser. To simulate this trick with our container, we would have to somehow cram all the marbles into the bottom without making the container any wider. Bosons can behave in strange ways.

The physics of Josephson junctions is yet another direct consequence of quantum mechanics. Josephson tunnel junctions are a manifestation of a more general phenomenon known as the *dc Josephson effect*. A simple theoretical experiment conveys the basic principle of this effect. Imagine that we have two identical chunks of superconductor sitting side-by-side and we attach one wire to each, hook the other ends up to a battery, and wait for current to flow. What would happen? Nothing, of course. Current can only flow in a complete circuit; the superconductors are not connected, so electricity cannot flow. Next we fuse the two chunks together into one piece and try the same experiment. Now a supercurrent—a zero-resistance current—can flow, since we have only one piece of superconductor. Finally we try a compromise experiment. We don't fuse the superconductors but we get them extraordinarily close together. What happens this time? If we had only classical physics to work with, nothing would happen, just as when the superconductors were far apart. Classically, two superconductors brought very close together are no better than two superconductors a foot apart at carrying current between them. However, quantum mechanics saves the day once again. Just as in the tunneling process for normal electrons, the superconducting electrons can manage to get across the "no-man's land" between the two superconductors and sustain current flow. The two superconductors join forces for the purposes of allowing current to flow. For sufficiently close superconductors, the electrons behave as though a single piece of superconductor was present. A Josephson tunnel junction consists of two pieces of superconductor separated by an extraordinarily thin insulator. But unlike an ordinary superconducting tunnel junction, a Josephson junction permits the flow of current with no voltage. The Josephson effect ensures that the combination will behave like a single superconductor; it will pass current with zero resistance.

There is one fly in the ointment, however. Two superconductors that join forces because of the Josephson effect make for a very weak superconductor, one that can't carry very much current or withstand very high magnetic fields. If we think of a superconductor as a chain in which electrical current is passed from link to link, a Josephson junction acts like a weak link in the chain. While the individual superconductors that form a Josephson junction might be able to carry rather substantial supercurrents, the Josephson junction itself will carry only a fraction of that current. Such a weak link makes for a useful device because it gives us a way to alter the properties of a superconducting circuit with far smaller currents and fields than the rest of the circuit requires. As we will see, weak links and Josephson junctions are the main building blocks of superconductive electronics.

How does a Josephson junction behave in terms of current–volt-

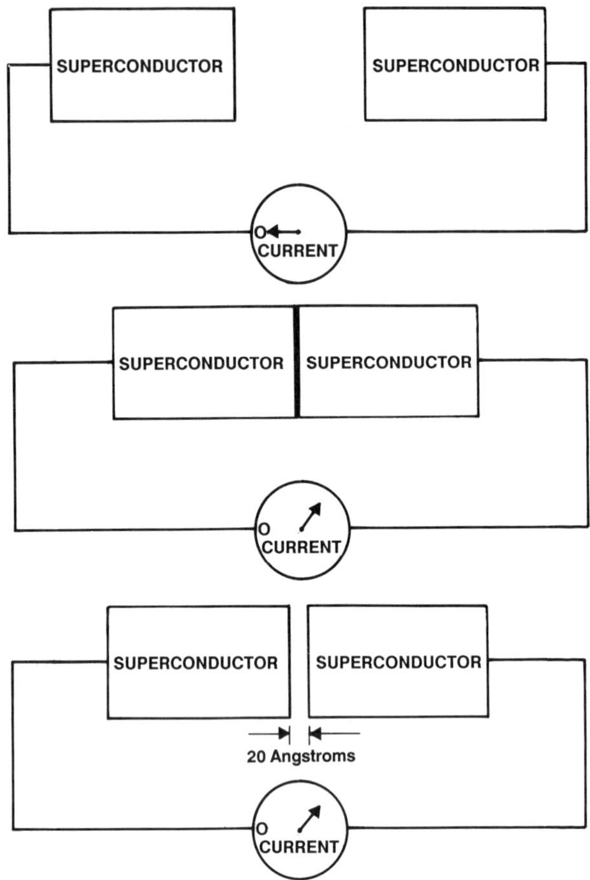

Josephson tunneling can occur between closely-spaced superconductors.

age characteristics? An ordinary superconducting tunnel junction cannot pass any current at all at zero or low voltage. The Josephson junction, on the other hand, passes current with no voltage up to some maximum amount—called the *Josephson critical current*—and then the effect goes away. The two superconductors abandon their alliance and supercurrent can no longer flow. At that point, we go back to the usual superconducting tunnel junction behavior (only normal electrons can tunnel through the barrier, which requires a voltage). When we add this additional behavior to our current–volt-

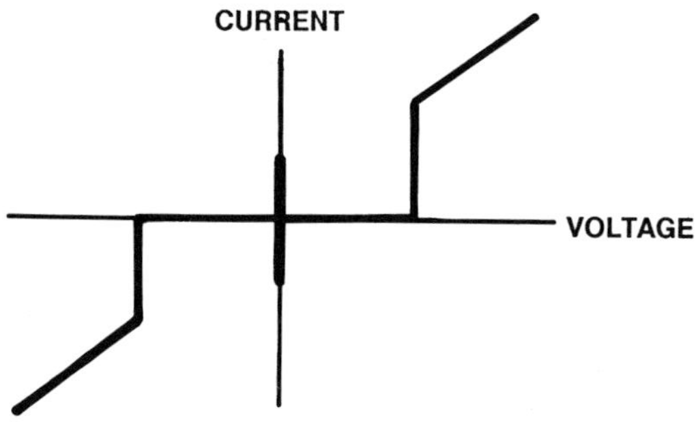

Josephson junction current–voltage characteristic.

age characteristic, we find that our nonlinear device has gotten even more nonlinear; it looks even less like a resistor. In fact, it now behaves like something called a *bi-stable device*; knowing the current across the junction no longer tells us what the voltage is. There are two different answers depending on the junction's electrical history. When a given amount of current flows, there can either be no voltage across the junction—when the Josephson supercurrent flows—or there can be a finite voltage—when it behaves like a regular superconducting junction. Such a thing has tremendous device potential if there exists a way to switch from one kind of behavior to another. Fortunately for people who want to use Josephson junctions, not only can we switch between the two voltage modes but we can do it extremely rapidly. High-speed electronic switches are the heart of any computer, and later we will see how Josephson junctions may someday be used in the fastest computer ever imagined.

The Josephson effect that we have discussed is called the dc Josephson effect. Brian Josephson predicted another phenomenon as well, one now called the *ac Josephson effect*. The effect comes about by a simple physical process. As we have seen, a Josephson junction can sustain current without any voltage. The junction requires no energy to pass current because a supercurrent can flow across the insulating barrier. The ac Josephson effect occurs when we intentionally apply a voltage across a Josephson junction.

Applying voltage adds energy to the electrons in the junction, energy that the electrons cannot use. In order to rid themselves of this excess energy, the electrons must radiate it away. According to electromagnetic theory, in order to radiate energy, electrons must undergo constant acceleration. They must always either be speeding up or slowing down. The electrons in a junction accomplish this by oscillating back and forth across the barrier. In other words, an alternating current will flow. The electrons then radiate energy at the frequency of the ac current. The voltage across the junction determines the frequency of the current; even a millionth of a volt will produce currents that oscillate at half a billion cycles per second. In the simplest terms, in the presence of a voltage, a Josephson junction becomes a tiny radio transmitter.

Thus we see that the physically simple Josephson junction exhibits two different kinds of unusual electrical behavior as a result of unique cooperative behavior among superconducting electrons. We haven't yet said anything about why two superconductors cooperate with one another to form a Josephson junction or, for that matter, what role quantum mechanics plays in the process. The answer to both questions comes from wave mechanics and a concept called *long-range order*.

Long-Range Order and the Flux Quantum

Long-range order means organized behavior over a large area. In superconductors, we are talking about something that allows the superconducting electrons to behave in an aggregate fashion rather than as independent particles. Just what does this mean? Electrons can be described as waves. As we know, waves have two fundamental properties: amplitude and frequency. The amplitude tells us how strong the wave is and the frequency tells us how often to expect a crest or a trough to come by. But to fully describe the wave at a given time, we need to know its phase. The phase of the wave tells us where we are on the wave, whether trough or crest or in between. Phase is not a new idea to us. Many phenomena that are periodic or cyclic in nature have phases. The most familiar example is the phases of the moon.

We can fully describe a wave by its amplitude and its phase. By fully describe, we mean that if we know the amplitude and phase of a wave of a certain frequency at a given time, we know what it will be at any other time as well. This is true for ocean waves as well as for electron waves. In fact, the primary task in quantum mechanics is to calculate the wavefunction for a given object, which tells us its amplitude and its phase. The wavefunction corresponds to a mathematical description of the wavelike nature of an object. If we know the wavefunction for an object, we can predict its physical behavior over the course of time.

With this picture, we can now understand a fundamental distinction between normal electrons and superconducting electrons. Normal electrons have wavefunctions with amplitudes and phases that are essentially independent of those of other electrons. If we know the wavefunction for electron A in a metal, it tells us nothing about what electron B will do. Electron B has its own distinct wavefunction. In contrast, superconductivity results from electrons cooperating with one another in such a way that they share a common wavefunction. If we know the amplitude and phase for electron A, we even know the amplitude and phase for electron Z. We call it long-range order when the amplitude and phase at one place tells us about the amplitude and phase far away. In this case, the whole ensemble of superconducting electrons can be described by one wavefunction. A single wavefunction for all the electrons is precisely what we mean by the term *macroscopic quantum mechanics*. When so many particles occupy the same quantum state, we can directly observe quantum mechanical effects on a macroscopic level. In fact, most of the unique properties of superconductors are consequences of macroscopic quantum mechanics.

The Josephson effect in particular results from two superconductors acting to preserve their long-range order across an insulating barrier. With a thin enough barrier, the phase of the electron wavefunction in one superconductor maintains a fixed relationship with the phase of the wavefunction in another superconductor. This linking up of phase is called *phase coherence*. It occurs throughout a single superconductor, and it occurs between the superconductors in a Josephson junction. Phase coherence—or long-range order—is the essence of the Josephson effect.

The last quantum effect in superconductivity that we will talk about is one that ties together many of the other phenomena we have discussed. It involves zero resistance, diamagnetism, quantum mechanics, and long-range order. Another superconducting experiment reveals this effect.

Suppose we place a ring of metal in a magnetic field and cool it down below its superconducting transition. What happens? We know that the superconductor will expel the field from its interior because of the Meissner effect. But something else happens as well. A current starts flowing in the ring. Why? The current flows because of yet another quantized property in nature: magnetic flux.

Magnetic flux is related to magnetic fields. It provides a measure of how large an area a given magnetic field covers. The flux tells us the total amount of magnetic field rather than just the concentration of the field. The comparison corresponds to the difference between a pouring rainstorm that covers a city block and one that covers a square mile. The strength of the rain might be the same in both cases but the square-mile storm has more rain. The flux of water would be the rate at which it falls times the area it covers. Similarly, magnetic flux corresponds to the field strength times the area it covers.

We will see that the superconducting ring experiment demonstrates the quantization of magnetic flux; flux can only exist in a superconducting ring in discrete units. The quantization of flux in a ring turns out to be fundamental to the nature of superconductivity. How so?

Consider the superconducting wavefunction for the electrons in a ring of superconductor. Suppose we know the amplitude and phase of the wavefunction at one spot on the ring. According to what we said about long-range order in a superconductor, we ought to know the amplitude and phase at any other spot on the ring. In fact, we have to know the amplitude and phase all the way around until we get back to where we started.

Now we apply a magnetic field to the ring. What happens? We already know that the superconductor expels the field from its interior because of the Meissner effect. But something else happens as well. Magnetic fields cause the phase of a wavefunction to change. We can't understand this intuitively; the result comes out of the nature of magnetic fields and the theory of quantum mechanics. If we know a given wavefunction, we can calculate what it will be in the presence of a magnetic field. In a superconducting ring, the magnetic field causes the phase of the wavefunction to change as we move around the ring. This should not trouble us; we simply determine new values for the phase of the wavefunction. But suddenly, we get into trouble. When we get all the way around the ring, we are back at our starting position and the phase has to be the same as it was when we started. But it can't be, because the magnetic field has changed it.

Think of a bumpy racecourse that has hills and dips that alternate

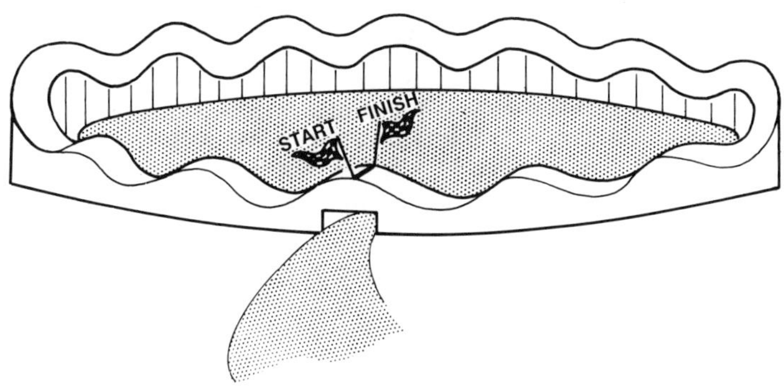

A bumpy racecourse representing the wavefunction in a superconducting ring.

as we go around it. These undulations are like the wavefunction of the superconducting electrons in a ring. Suppose we place the starting line at the top of a hill. From there, we proceed around the course. We go down and up and down and up, following the terrain. We get to the finish line—back where we started—and where must we be? Back on the original hill, of course. If the finish line was now at a dip, there would be something really wrong. It is the same place we started from, after all. In the same way, the ordered superconducting wavefunction demands that we end up with the same phase we started with when we go around a ring. Why is this important?

Because magnetic fields alter the phase of a wavefunction, the phase of the superconducting wavefunction in our ring in a magnetic field has to change from point to point as we go around the ring. Therefore, something has to give: either the long-range order of the wavefunction or the magnetic field. If superconductivity is to persist, then what gives must be the magnetic field. The field needs to be different, so the superconductor changes it by flowing current around the ring, creating an opposing field. How much current must flow? The obvious answer to guess is that there has to be enough to generate a magnetic field that exactly cancels out the applied field and leaves the wavefunction's phase completely undisturbed. However, the real answer is that the superconductor can cheat a little.

In order for the wavefunction to retain its order and to avoid any paradoxical results, the phase only has to end up where it started when we go all the way around the ring. How it gets there doesn't matter. In terms of our racecourse, we can add as many hills and dips as we'd like as long as we finish with the same hill. The result of this loophole in wavefunction accounting is that we don't need to get rid of all the magnetic flux threading a superconducting ring. We only need to make sure that it is the right amount to keep the phase of the wavefunction from getting out of sync with itself.

So, when we apply a magnetic field to the ring, enough current flows to change the flux in the loop to a legal amount, not too large and not too small. Only a discrete set of flux values are permitted that allow the wavefunction to rejoin its own end. This is the origin of flux quantization in a superconducting ring. We constantly refer to the less familiar concept of flux rather than magnetic field because flux and not field is the quantized property. Keep in mind that the two are intimately related; the flux in a ring is just the magnetic field times the area enclosed by the ring. Physicists call the flux contained in a super-conducting ring (plus an additional contribution from the current in the superconductor that is usually negligible) the fluxoid. The fluxoid comes in extremely small quantized units called *fluxons*. The size of a fluxon can be calculated from fundamental physical constants and corresponds to about a millionth of the earth's magnetic field over a square centimeter, a tiny amount indeed.

This whole description seems to be the most far-fetched topic we have yet covered. Why should we care about it at all?

The answer is that flux quantization provides the operating mechanism for SQUIDs (Superconducting Quantum Interference Devices), which, as we have said, have numerous applications in science, medicine, and high-performance electronics. SQUIDs take advantage of the flux quantization in a superconducting loop to produce an incredibly sensitive detector of small magnetic fields, and in turn, of small currents and voltages. In a SQUID, the loop contains a Josephson junction that acts like a switch—a weak link that can open and close the superconducting circuit—while flux quantization allows precise amounts of current to be stored in the loop. We will have much more to say about SQUIDs later on.

More fundamentally, flux quantization gives us an important insight into why superconductors carry current with no resistance. When we apply a magnetic field to a superconducting ring, a current starts to flow. This is simply a diamagnetic response. If it was flowing in an ordinary conductor, the current would quickly die out from resistive losses. In the superconductor, on the other hand, there are two choices: either the long-range order disappears entirely or the current has to fall off in discrete steps. Such steps—known as quantum jumps—would always result in a current that generates an allowed amount of flux in the ring. If the current dropped down by a smaller amount than one of these steps, the remaining current in the ring would generate a fractional number of flux quanta in the ring; there would then be a mismatch in the phase of the wave function. Such a quantum jump from one amount of kinetic energy to another would be perfectly possible for a single electron to do; it would simply drop into a lower energy state. On the other hand, reducing the current in a ring requires billions of billions of electrons to reduce their energy. To maintain long-range order, all the electrons would have to do it simultaneously. It is fantastically unlikely for the whole ensemble of superconducting electrons simultaneously to lose the same amount of energy in this way. Things that are fantastically unlikely according to quantum mechanics are things that we can safely say will never happen. If the electrons cannot drop into a lower energy state, the only way they can preserve their long-range order is to sustain the current in the loop. Thus we get resistanceless current.

We are beginning to understand how the long-range order in superconducting electrons leads to the various phenomena in superconductors. It is certainly fair game to ask two important questions: how does long-range order come about in the first place and why should the electrons want to preserve it? The answers are embodied in the BCS theory that we take up shortly.

By now we have covered nearly all of the fundamental elements of superconductivity. When we start discussing the broad variety of applications of superconductivity, we will see that all the phenomena we have discussed can be exploited in a wide variety of devices, machines, and systems, both practical and fantastic.

At this point, however, we return to the early years of the twentieth century and see how superconductivity came to be discovered, investigated, and finally understood.

CHAPTER 4

The Era of Discovery

Superconductivity was the kind of unexpected discovery that scientists dream of all their lives. No one was looking for it, but the first person who even had a chance at finding it, did. No flash of insight or stroke of genius was needed. Instead the discovery required many years of hard work and tremendous progress in cryogenics, the science of refrigeration. As we will see, the early history of superconductivity and the history of cryogenics are inexorably intertwined.

Making things cold was what was needed. Nowadays we take refrigeration for granted. We buy our sturdy side-by-side at the appliance store, plug it in, and our food stays cold for years to come. But in the nineteenth century, pioneers in the new science of cryogenics worked hard to perfect refrigeration techniques. They reached ever-lower temperatures with their ingenious machines that condensed gases into the liquid state. As the century drew to a close, refrigeration experts moved toward the goal of reaching the very bottom of the absolute temperature scale.

For most of us, a refrigerator is a kitchen appliance that cools our food. But the marvelous machines that turn air into a fuming bluish liquid 200 degrees below the freezing point of water are also refrigerators. They all work on the principle of converting the heat contained in a substance into mechanical work. We can easily demonstrate this form of energy conversion. Cup your hands in front of your mouth and exhale, keeping your mouth wide open. Warm air blows onto your hands. Now exhale again, this time with your lips tightly pursed. The air is suddenly much cooler. Your lips act as a nozzle through which you force air at high speed. Once outside, the air expands and cools.

Several other phenomena are probably also at work in this little demonstration, but in any case, the particular effect of the cooling of a compressed gas when it is allowed to expand is central to the operation of many refrigerating machines. In order to liquefy a gas, we first compress it in the presence of a cool thermal bath. We then allow the gas to push against a piston or to expand into a vacuum, thus expending energy. The gas then cools further. By repeating the process, we can eventually lower the temperature enough to liquefy the gas.

The lowest temperature a refrigerator can reach depends on its working gas. We know that water freezes into ice at 32°F (273 K). Thus the easiest way to cool something to that temperature is to store it in ice. Carbon dioxide freezes (into dry ice) at −108°F (195 K), giving us a way to sustain far lower temperatures. In the same way, the characteristic freezing or condensing temperatures of other gases provide convenient ways of reaching low temperatures. To reach very low temperatures we must remove heat from a specimen; immersing it in a liquefied gas removes heat as the liquid boils.

Race for the Cold

Scientists perfected the techniques for condensing various gases throughout the nineteenth century. Each new gas and each lower temperature required more sophisticated refrigeration machines. Michael Faraday froze carbon dioxide in 1845. From that point on, the challenge was far greater but scientists around the world took it on. Raoul Pictet in Switzerland and Louis Paul Cailletet in France independently managed to produce droplets of liquid oxygen and nitrogen, the primary components of air, by 1877. They had thereby succeeded in reaching temperatures as low as 77 K. In 1883, the Polish scientist Zygmunt Wroblewski produced a "quietly boiling" test tube of liquid oxygen. Of all the so-called permanent gases—elements that exist on earth only in the gaseous state—there were only hydrogen and helium left to be liquefied.

There are two main difficulties with really low temperatures: how to get them and how to keep them. Before the first problem could be tackled any further, the second had to be solved. The tiny droplets of liquid air produced in the 1870s would rapidly boil away in their container. The heat coming in from the outside world was simply too great for the cold liquid. An effective means of insulation was needed.

James Dewar, a Scottish physicist, provided the solution in 1892. His invention was a silver-coated glass vessel with double walls and an evacuated space between them. It provides extremely good thermal insulation from the warm outside world; even liquid air could last for hours in one of Dewar's vessels. Such containers for cold liquids are called Dewar flasks, or "dewars" by scientists to this day. The rest of us call them by the trademark name "Thermos." With such flasks, scientists could transport, store, and make use of liquefied gases at extremely low temperatures.

A worldwide race was on to liquefy the remaining gases and reach ever-lower temperatures. Among the participants were Olskewski in Poland, Dewar at the Royal Institution of London, and Heike Kamerlingh Onnes at the University of Leiden. Dewar ended up being first to liquefy hydrogen in 1898. By doing so, he reached the remarkable temperature of 20 K and, a year later, even produced solid hydrogen at 16 K. Only helium now remained to be liquefied.

Until the turn of the twentieth century, nobody had even considered liquefying helium. The lightweight gas was first discovered in the spectrum of the sun in 1867. And no one had found any trace of it on the earth until William Ramsey did so in 1895. Thus, the necessary know-how for liquefying helium evolved about the same time that helium itself became known. The race for the lowest temperatures continued with Olskewski, Dewar, Onnes, and now Ramsey. But Onnes had a big advantage over the others: he was able to get a supply of helium gas.

Onnes had been appointed to the chair of experimental physics at the University of Leiden in 1882 at the age of 28. He would hold that position for 42 years. From the very beginning he established an ambitious program of research on the properties of gases. The work was founded on the theories of his mentor Johannes van der Waals, who had developed a theory called the Law of Corresponding States that sought to explain the behavior of gases. To confirm the predictions of van der Waals' theory, Onnes needed to study gases over as wide a temperature range as possible, so it was advantageous to make use of gases that liquefied at very low temperatures.

Years of careful planning and systematic expansion of Onnes' laboratory paved the way for his success. By 1894, he had constructed large-scale liquefaction plants for air. By 1906, he had a very successful hydrogen liquefier in operation, a major improvement over Dewar's first machine. A year later, Onnes calculated that the liquefying temperature for helium would be about 5 degrees above absolute zero. He designed and built the multistage apparatus necessary for the task. The biggest problem was obtaining a supply of helium gas.

Scientists call helium an inert gas; it forms no chemical compounds. It is also extremely light and tends to escape from the atmosphere into outer space. We are most familiar with its use for lighter-than-air balloons. The element's inertness and lightness result in the almost total lack of free helium on the earth. The atmosphere contains only about one part in 200,000 of helium, which explains why it took so long to be discovered on our planet. The gas is produced as a by-product when uranium decomposes during its natural process of radioactive decay after which the helium sometimes lodges in the surrounding rocks. Most of the world's supply of helium comes from natural gas wells in the United States, which are particularly good sites for trapped gases. It also lies trapped in certain other mineral deposits, including some uranium salts and a brownish crystalline sand called monazite. Kamerlingh Onnes was fortunate to have a brother at the Office of Commercial Intelligence in Amsterdam, who succeeded in locating a supply of monazite sand in North Carolina. After a tedious process of reaction and purification, Onnes managed to produce the rather sizable supply of helium gas he needed. One must realize that it takes 700 liters of helium gas in order to produce a liter of the precious and dramatically denser liquid.

The helium gas itself needed to be painstakingly purified for the liquefaction process. Any traces of other gases would produce frozen crystals that would clog the fine tubing in the liquefier. Liquid hydrogen was needed to precool the helium to a low enough temperature for the final refrigeration stage. The experiment began at 5:30 on the morning of July 10, 1908. Thirteen hours later came the moment Onnes was waiting for. He later remarked "I was overjoyed when I could show liquefied helium to my friend van der Waals, whose theory had been my guide in the liquefaction up to the end." Helium liquefied at 4 degrees above absolute zero. Kamerlingh Onnes could now fully pursue his studies of the properties of gases.

Over the next year or two, he tried to produce solid helium by pumping on the liquid to reduce its temperature. Despite the fact that

H. Kamerlingh Onnes, discoverer of superconductivity beside his helium liquifier. (Courtesy AIP Niels Bohr Library, Rijksmuseum roor de Geschiedenis der Natuurwetenschappen to Leiden)

he reached the extraordinary low temperature of 1 K in these attempts, Onnes could not get helium to solidify (and, in fact, it never does without the application of tremendous pressure). Wisely, he decided to pursue another avenue of research: The electrical resistance of metals as a function of temperature.

Onnes chose this area of research because at the turn of the century, the low temperature behavior of metallic resistance was a very controversial subject. Nobody was looking for superconductivity, they were just trying to see what happens to resistance when metals get very cold. No less than three competing theories were championed by leading scientists of the day. Matthiessen predicted in

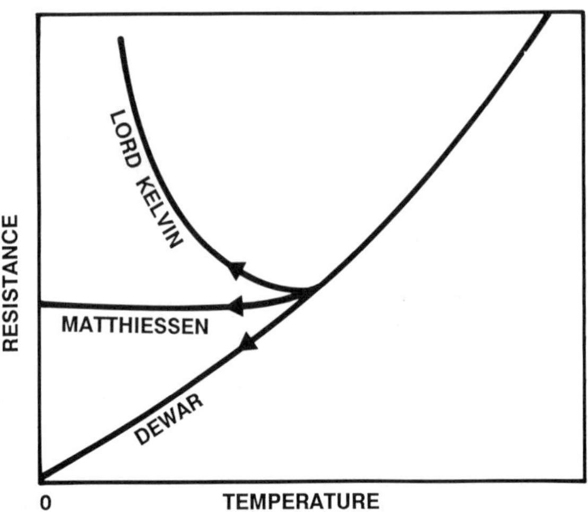

Three predictions of metallic resistance behavior near absolute zero.

1864 that the resistance of metals would be constant at low temperatures because impurities dominate it. Lord Kelvin, for whom the absolute temperature scale is named, predicted that the resistance of a metal would sharply increase at the lowest temperatures. His 1902 theory claimed that the itinerant electrons in a metal would become frozen to their parent atoms at the lowest temperatures and electrical conduction would therefore cease. James Dewar, on the other hand, predicted that the resistance of a metal would gradually vanish at low temperatures, as the thermal vibrations contributing to resistance disappeared.

The Leiden group favored Kelvin's theory. Dewar's own data on gold and platinum couldn't settle the debate. At the lowest temperatures Dewar could produce (16 K), it was unclear just which way the resistance would end up, although it did seem to flatten out in accord with Matthiessen's rule. Onnes and Jacob Clay repeated Dewar's experiments at lower temperatures. Again, no sign of the expected upturn was seen. In fact, Onnes at first decided that the resistance of gold actually dropped to zero at liquid helium temperatures. But this was not the discovery of superconductivity, it was merely the scientists' inability to detect the tiny resistance of gold at such low temperatures.

The Discovery

In late 1910, Onnes began another series of resistance measurements, collaborating with Cornelis Dorsman and Gilles Holst. This time they used more sophisticated measurement techniques. They could now detect exceedingly small resistances. The researchers carefully investigated the effects of impurities on the low-temperature resistance of gold and platinum, concluding that in both metals it is dominated by a residual resistance due to the presence of the impurities.

To settle the issue once and for all, they needed a more perfect sample of a metal. Mercury, a liquid at room temperature, provided the answer. With care, one can distill it to amazing purity. Producing fine mercury wires was a painstaking task of filling extremely fine glass capillary tubes with the liquid metal and then freezing them. They used a new and improved experimental setup to measure the mercury resistance samples.

A brief announcement to the Royal Netherlands Academy in Amsterdam on April 28, 1911, reported ("with all reserve") the initial results. At a temperature just above 4 K, the resistance of mercury had apparently vanished. The name of Kamerlingh Onnes would go down in history.

An interesting aside to the story is that, in fact, Gilles Holst actually made the historic measurements. His name has become lost in the recesses of history, as is often the case with junior researchers working under a famous scientist. In fairness, the experiments at Leiden were clearly designed and the research led by Kamerlingh Onnes. While Holst conducted most of the superconductivity experiments at Leiden until his departure in 1914, it was simply not the custom of the time to include junior staff as co-authors. Onnes did acknowledge the assistance of Holst in his articles and later recommended him for membership in the Royal Netherlands Academy.

In any event, throughout 1911 they performed the experiment again and again. Holst checked and rechecked his instruments. Gradually Onnes realized the significance of the result. Describing the remarkable change in mercury's resistance in a *Communication* of the Leiden Laboratory, Onnes wrote: "At this point within some hundredths of a degree came a sudden fall, not foreseen by the vibrator theory of resistance that I had framed, bringing the resistance at once to less than a millionth of its original value at the melting point and a thousand millionth of it at the lowest temperatures obtained. Mercury has passed into a new state, which on account of its extraordinary electrical properties may be called the superconductive state."

Kamerlingh Onnes was not just lucky to discover superconductivity. Careful work and planning allowed all the pieces of the puzzle to come together. Hydrogen was liquefied, the Dewar flask was developed, helium was discovered on the earth, and, with Onnes' efforts, was finally liquefied. The time had come for the discovery.

At Leiden, the remaining years before World War I were filled with important discoveries about the new superconducting state. Kamerlingh Onnes was an exceedingly careful experimentalist and he expended great effort to improve upon the early measurements of superconductivity. A long series of communications from Onnes' laboratory documents the discoveries made over the following 3 years.

Onnes investigated the superconducting transition itself to find out how low the resistance really was and over how small a temperature interval the transition occurred. The early measurements showed that mercury's resistance had dropped to no more than a ten-millionth of its room temperature value. By December of 1911, Onnes had lowered this upper estimate to a billionth of the original resistance. By 1913, he was confident that the resistance of the superconductor was less than a ten-billionth of the normal state value. It seemed clear that the resistance of mercury was zero in the superconducting state.

The interval in temperature between the normal state and the zero resistance state is called the transition width. Onnes' best measurements determined the width of the transition in mercury to be about 0.02 K. He suspected, however, that the transition to the superconducting state was instantaneous (meaning that no width in temperature was involved), but he could not demonstrate it. Later work showed that the transition width depends on the purity of the material. The very best superconductors prepared today have transition widths of hundred-thousandths of a degree.

In the early years, superconductivity experiments were tedious and difficult. It took many hours to produce the necessary liquid helium for the measurements and preparing the delicate capillary tubes of ultrapure mercury was a constant source of frustration as the tiny filaments of the frozen metal would often break. Two discoveries at the end of 1912 drastically changed the experimental situation.

The first occurred when the experimenters deliberately added traces of other metals to the mercury. Amazingly, the superconductivity was unaffected. Onnes wryly regretted all the time and effort spent in distilling mercury that could have been saved.

The second discovery came on December 3, 1912. Onnes performed resistance measurements on tin and found that it too loses its resistance at low temperature (about 3.7 K in this case). Soon thereafter, the Leiden group found superconductivity in lead above 6 K. The days of working with hard-to-handle mercury were over.

Now Onnes was able to do a variety of experiments on the new superconductors and soon produced the first superconducting coils. He immediately recognized the practical potential of powerful electromagnets made from resistanceless wire. Unfortunately, his dream of such magnets would not be realized in his lifetime because the early superconductors could not sustain high enough magnetic fields. Later discoveries, of course, would vindicate his vision.

Early on, Onnes discovered the critical current effect in superconductors, observing that the remarkable zero-resistance state would disappear with the application of a sufficient amount of current. He called this phenomenon the "threshold value" for the current and did a series of experiments that demonstrated the way the critical current changes with temperature. These studies of critical currents occupied the first half of 1913. On December 10 of that year, the Swedish Academy awarded Onnes the Nobel Prize in physics "for his investigations on the properties of substances at low temperatures."

The Leiden group spent the last year before the war studying persistent currents in coils of lead with ends fused together. They were able to induce currents in the coils that flowed for hours on end with no measurable reduction. Many scientists came to Leiden to witness the miracle. Unfortunately, the outbreak of war brought a halt to experiments with liquid helium; the rare gas could no longer be imported from the United States.

By the middle of the decade, scientists already knew quite a bit about superconductors. Onnes had observed the critical magnetic field and Francis Silsbee at the National Bureau of Standards in Washington correctly deduced its relationship to critical currents. For the superconductors that Onnes studied, the critical current was the amount of current necessary to induce a magnetic field equal to the critical field. Thus the two phenomena were really only one. Such discoveries represented real scientific progress. But the major problem with superconductivity remained the inability of scientists to explain how it worked. Accepted theories simply could not handle the new phenomenon.

Kamerlingh Onnes himself made little progress in explaining his results. At one point he suggested that perhaps the entire electron theory of conduction had to be revamped in light of superconductivity. The top scientific minds of the day took a crack at explaining the Leiden results. Frederick Lindemann, an English physicist famous for a theory explaining how solids melt, proposed a theory in which a lattice made of electrons drifted en masse through the crystal lattice without any hindrance from the ions. Sir Joseph Thomson, who discovered the electron, proposed a complicated explanation based on metal atoms composed of charged dipoles that align in an electric field. Perhaps the most courageous view was expressed by the founder of quantum mechanics, Max Planck, who was honest enough to admit that he simply had no ideas about the mechanism behind Onnes' discovery.

The situation was best summed up by Albert Einstein, who said in 1922, "with our considerable ignorance of complicated quantum-mechanical systems we are far from being able to formulate these ideas in a comprehensive theory. We can only attack the problem experimentally." And it was truly the experimentalists who had more ground-breaking work to do. Before they could explain the behavior of superconductors, scientists had much more to learn.

For another decade or so, Leiden remained the world center for superconducting discoveries. Kamerlingh Onnes—respectfully given the title "the gentleman of absolute zero" by his friends—died in 1926, but not without leaving behind a legacy in Leiden. It remained an institution dedicated to the motto he first proclaimed at his inaugural address in 1882—*Door meten tot weten* (comprehension through measurement). Onnes' passing marked the end of an era in superconductivity—in some sense, its childhood.

Growth of the New Science

The adolescent period for superconductivity—when scientists' naive view of the phenomenon gave way to a more sophisticated understanding—also started in Leiden. The dominance of the Dutch in low-temperature physics was no coincidence. Until the end of World War II, there were very few laboratories in the world that were equipped to study superconductivity. The work finally spread elsewhere when new helium liquefiers were developed by Peter Kapitza and by Samuel Collins, who produced the first commercial apparatus. Until these developments, discovery after discovery continued to come from Holland.

In 1931, W. J. de Haas, the new co-director along with W. H. Keesom of the Kamerlingh Onnes Laboratory, discovered superconductivity in alloys—combinations of metals. In due course, alloys would be a key to the development of both scientific and technological advances in superconductivity. At the same time, scientists continued to measure various properties of superconductors, looking for other changes besides the cessation of resistance. Most properties of metals seemed to be unaffected by the onset of superconductivity, but Keesom and J. A. Kok observed an abrupt change in the specific heat of tin at its superconducting transition temperature. When the specific heat of a substance—which measures the energy required to raise its temperature—suddenly changes, it is the signature of a phase transition. We are familiar with other phase transitions, such as the freezing of water or the melting of ice. But the idea of superconductivity as a phase transition shed a whole new light on the subject. The loss of resistance in a metal might only be an artifact of a more fundamental change in the state of the material. The next important discovery in superconductivity confirmed this conclusion.

For over 20 years superconductivity had been synonymous with the zero-resistance state. If superconductors were simple perfect conductors with no other essential properties, then according to the predictions of classical electrodynamics, the onset of superconductivity would trap any existing magnetic field within the material. Maxwell's electrodynamics theory requires that the magnetic field within a perfect conductor cannot change; it does not require that the field is zero. Early magnetic experiments by Onnes agreed with these predictions. What went wrong? The group at Leiden had made the mistake of using a hollow sphere of lead for their measurements. Such an object does not demonstrate the Meissner effect—the expulsion of magnetic fields—because the hollow space is not a superconductor. Thus, while no magnetic field could get inside when such a sphere was already superconducting, it would be trapped inside if the superconductivity was turned on afterwards. The superconducting shell could only shield out magnetic fields, not expel them.

The Leiden stranglehold on superconductivity discoveries was finally broken by a Berlin laboratory in 1933. Here Walther Meissner and his colleague Robert Ochsenfeld performed their landmark experiment that demonstrated the expulsion of a magnetic field from the interior of a metal cooled below its superconducting transition. A perfect conductor does not allow magnetic fields to enter it; a superconductor also kicks out whatever field was there before it was superconducting. The classical theory of electromagnetism could not predict this perfect diamagnetism. It was now clear that superconductors were not just perfect conductors.

The biggest impact of Meissner's result was that it made superconductivity fair game for the discipline of thermodynamics. Thermodynamics is the science that relates states of matter—like liquid, solid, magnetized, or superconducting—to heat and other forms of energy. The task of thermodynamics is to determine how the state of a system changes with variations in temperature, volume, pressure, magnetic field, or other pertinent parameters. The formalism of thermodynamics allows the state of a system to be determined by specifying all the forms of energy in the system in terms of these parameters. This may sound rather abstract, but it provides a very powerful theoretical tool for understanding such ordinary phenomena as the freezing and boiling of water.

Thermodynamics can successfully predict the properties of a phase transition—like the transition from liquid water to gaseous steam—if the transition is reversible between good thermodynamic states. Reversible means that it is possible to go back and forth between the two phases with nothing changing but the parameter we are varying. For example, we lower the temperature of water and get ice; we raise the temperature of the ice and we get water. Liquid and solid are the phases and temperature is the parameter; nothing else changes. Most important, we can't tell the difference between warm water that used to be ice and water that was always warm. Because of this, liquid and solid are good thermodynamic states.

Until the discovery of the Meissner effect, scientists did not believe superconductivity was a good thermodynamic state. If superconductors would trap in a magnetic field that was applied *before* cooling but exclude a magnetic field applied *after* cooling, superconductivity could not be a good thermodynamic state. In either case, we would end up with the same superconductor sitting in a certain magnetic field but we would be able to tell the difference between the superconductor cooled in a field and the superconductor for which

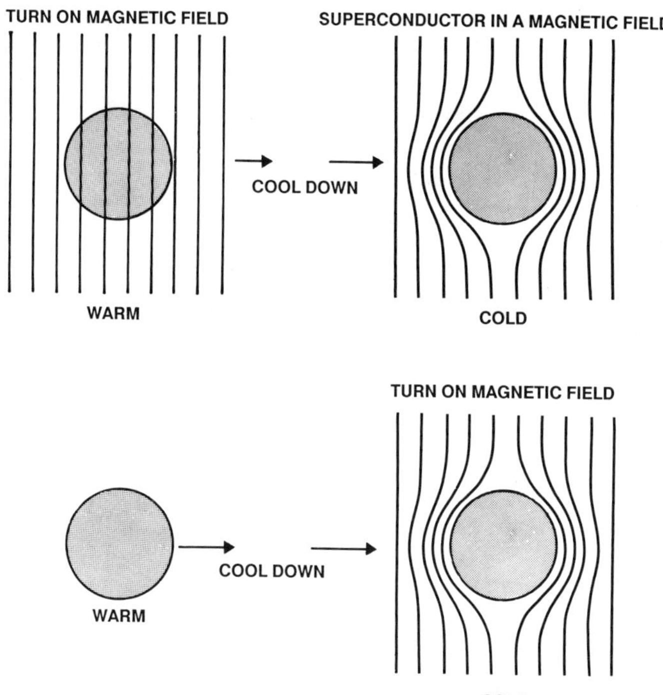

The Meissner effect—a superconductor in a magnetic field.

the field was applied later. This would be just like previously frozen water being different in some way from ordinary water. The discovery of the reversible Meissner effect eliminated this possibility and therefore demonstrated that thermodynamics could properly be used to describe superconductivity.

The discovery of the Meissner effect sparked a new flurry of superconductivity research. The nature of the thermodynamic phase transition in superconductors immediately became the subject of great interest. Physicists categorize the freezing and boiling of water as *first-order phase transitions*. Such a phase transition always involves the presence of a *latent heat*, which is a quantity of energy necessary to complete a phase transition at the critical temperature. The extra heat that must be added to water when it reaches 212°F before it will boil provides a familiar example. Boiling, which requires this latent heat, therefore constitutes a first-order phase transition. Paul Ehrenfest, a theoretical physicist at Leiden, asserted that superconductivity in the absence of a magnetic field was in fact a different kind of phase transition, which he called a *second-order phase transition*. The hallmarks of such a transition are an abrupt change in the specific heat and the absence of a latent heat. There are now known to be a number of second-order phase transitions in nature, but superconductivity was the first to be identified. The brilliant Soviet theoretician Lev Landau worked out the details of an elegant theory of second-order phase transitions in 1937. This work would later be the basis for a very successful theory of superconductivity. A curious side note is that in the presence of a magnetic field, superconductivity becomes a first-order phase transition, complete with a latent heat.

The next clue about the thermodynamics of superconductors came from the measurements of critical magnetic fields. The destruction of superconductivity by the application of a magnetic field can be best understood by thermodynamic arguments about energy.

It is a fundamental rule of physics that physical systems seek their lowest possible energy state. It is for this reason that a ball will roll to the bottom of a hill and come to rest there; it has found the lowest state of gravitational energy. In order to think about superconductivity in this way, we can ascribe a certain energy to the superconducting state and a certain energy to the normal state. If at some temperature the superconducting state has lower energy than the normal state, then the superconducting state prevails. As the temperature is lowered further, the energy difference between the superconducting state and the normal state increases, making superconductivity even more favorable. This is the thermodynamic basis of the two-fluid model we talked about earlier.

Now we apply a magnetic field. What happens? If the superconductor is below its transition temperature, the Meissner effect occurs; the superconductor expels the field. Expelling the field requires some energy that must be supplied by the superconductor. The energy appears in the form of kinetic energy—energy of motion—of the electrons producing the magnetic screening currents. If too much magnetic field is applied, this added energy gives the superconducting state more energy than the normal state. Superconductivity is no longer the lowest energy state and is therefore destroyed.

This description of the critical field in superconductors agreed very well with the experimental results and formed the basis of the first successful model for the thermodynamics of superconductors:

the two-fluid model of Cornelius Gorter and H. B. G. Casimir, proposed in Leiden in 1935. In this model, the electrons in a superconductor come in two varieties: superconducting electrons and normal electrons. The electrons are like two interpenetrating liquids that fill up the metal—hence the name two-fluid model. The superconducting properties of the metal depend on how many of each kind of electrons are present. The more superconducting electrons there are, the stronger the superconductivity. The model specified a parameter for the strength of the superconductivity that is largest at very low temperatures and diminishes to zero as the temperature is increased to the critical temperature. All the parameters of the theory were chosen to fit experimental results. The two-fluid model enjoyed its greatest success with its connection to the work of a pair of German refugee scientists: the London brothers, whom we met earlier.

By the mid-1930s, Leiden finally took a back seat to the pioneering theoretical and experimental work taking place in England, Germany, the Soviet Union, and later, the United States. The political change in Germany brought about a migration of scientists, and in turn technology, in Europe. German refugees brought low-temperature technology to England early in the 1930s. The action shifted away from Holland by the middle of the decade when the fear of American helium finding its way into Nazi zeppelins shut off Leiden's supply of the gas until after the war.

The German refugee scientists who came to England brought small-scale helium liquefiers to such places as Oxford and Cambridge, and soon scientists were doing important low-temperature research at the British universities. Among the refugees were two brothers named Fritz and Heinz London. Fritz, the elder of the two, was a theoretical physicist and Heinz was an experimentalist. Heinz was intrigued by Meissner's paper of 1933 and immediately analyzed the thermodynamic consequences of the reversible Meissner effect. However, London was beaten to the punch by Gorter and Casimir's publication in a Dutch journal.

The next problem that the London brothers tackled was the electrodynamics of superconductors. Electrodynamics is the branch of physics that deals with the effects of electric and magnetic fields. It has its theoretical underpinnings in the work of James Clerk Maxwell who formulated a concise set of mathematical relationships expressing the interactions of electricity and magnetism. Despite their elegance, Maxwell's famous equations could not account for superconductivity.

Heinz London had worked on the problem of how electrons accelerate in a superconductor when he was still in Germany. With the publication of Meissner's results, Fritz continued where his brother left off and, in a flash of insight, conceived of a new pair of electrodynamic equations, first published in 1935. He realized that diamagnetism—the Meissner effect—was the key thing in superconductivity, not the zero resistance. London set down a new relation in which all supercurrents—the currents that flow in a superconductor—are determined by the local magnetic field rather than the electric field, as in an ordinary conductor. Scientists already understood that the magnetic field caused supercurrents to flow in the Meissner effect; London's first equation made the additional claim that even the current that flows in a superconducting wire is maintained by the magnetic field produced by it. His second equation expressed the Meissner result that the magnetic field inside a superconductor must vanish, not merely remain unchanged.

Why is the London theory important? Taken together the two London equations embodied a great deal of information about the superconducting state. The equations were an attempt to describe what was going on inside the superconductor, rather than merely describing the zero-resistance effects. Here at last was a theory that addressed the behavior of the superconducting electrons themselves. Of particular significance was the idea of the magnetic penetration depth. The London theory predicted that magnetic fields could not penetrate deep within the interior of a superconductor—this was the essence of the Meissner effect—but at the same time showed that magnetic fields could occupy a shallow region at the surface of the superconductor. The depth of this region was predicted by the theory, provided one knew enough about superconducting electrons.

The London penetration depth of a superconductor, as we now call it, constitutes one of two important characteristic lengths in a superconductor. In most materials, this length spans only a few hundred atomic layers. Thus a superconductor does not totally reject a magnitude turns out to be crucial for superconducting electronic shallow distance. Despite the shortness of the penetration depth, its magnitude will turn out to be crucial for superconducting electronic devices. According to the London theory, the actual magnitude of the penetration depth is determined by the mass, charge, and density of superconducting electrons. This last property makes contact with the two-fluid model of Gorter and Casimir. In that model, there is a parameter that indicates the strength of superconductivity in a sample. The Londons' density of superconducting electrons plays the same role. It vanishes at the transition temperature and reaches a maximum value at the lowest temperatures.

Between the two-fluid model and the London theory, a phenomenological picture of superconductivity was at hand. Scientists formulate phenomenological theories in such a way that they can be used to deduce observed behavior. As we have said, we call such theories phenomenological because they are not derived from more fundamental principles nor can they explain *why* the observed behavior occurs. This may seem like a nitpicking distinction, but it really isn't. A time-traveler from the seventeenth century might devise a phenomenological theory for electric lighting that asserts that the light is caused by pushing on switches. The theory isn't wrong but it doesn't explain how electric lights work. In the same way, a number of properties of superconductors make sense in the context of the theories of the 1930s, but these theories don't explain how superconductivity comes about.

Many other important developments in superconductivity took place in the mid-1930s. Fritz London continued to work on various aspects of his theory in Bristol. The Oxford group, led by Kurt Mendelssohn, investigated the curious occurrence of an incomplete Meissner effect in alloy samples. It appeared that only parts of the superconductor were excluding magnetic fields rather than the whole sample. Similar behavior was first discussed by Gorter and Casimir several years before in a paper predicting the occurrence of an "intermediate state" in samples of certain geometries in which rather than the entire sample becoming superconducting, there would be alternating regions of superconductor and normal metal. David Shoenberg's group at Cambridge, Shubnikov's group at Kharkov in the Soviet Union, and de Haas and others at Leiden all pursued the problem of superconductivity in alloys. It would be another 20 years before this nonideal superconducting behavior was fully understood. As we will see, such superconductors, now called Type II superconductors, provide all the commercially important materials for superconducting wire.

As the black cloud of World War II descended upon Europe, superconductivity research nearly ground to a halt. From about 1936 on, researchers seemed to desert the field. Some went on to pursue more promising scientific inquiries and some mysteriously disap-

Heinz London. (Courtesy AIP Niels Bohr Library, Photograph by Francis Simon)

peared in the political turmoil of the times, but most probably went off to work on radar and other technological needs of wartime. In any event, with a few exceptions, the development of superconductivity did not really resume until nearly 1950.

It was then that Fritz London made a daring prediction about the magnetic flux trapped inside a superconducting ring, which we discussed earlier. We define the flux in the ring as simply the product of the magnetic field strength and the area of the ring; London predicted that this flux could only exist in discrete units regardless of the strength of the original applied field. In other words, magnetic flux is quantized just like such properties as electrical charge and particle spin.

London had begun to think about the quantum-mechanical nature of superconductivity quite early on. He realized that the first

Fritz London. (Courtesy AIP Niels Bohr Library, Physics Today Collection)

London equation requires the superconducting electrons to have what he called "quantum structure on a macroscopic scale." By this he meant that there had to be the long-range order of the superconducting electron wavefunction that we talked about in the previous chapter. The superconducting electrons could not move independently, but rather could somehow only move in unison. In the language of quantum mechanics, this is known as a momentum-space condensation. London considered this property to be fundamental to the nature of superconductivity and recognized that it was the task of any real theory of superconductivity to explain it.

So in 1950, London predicted that magnetic flux would be quantized in a superconducting loop. At the time, this was a particularly revolutionary idea because it represented a macroscopically observable effect of quantum mechanics. Quantum effects had always been associated with the behavior of atoms, electrons, or subatomic particles, never before with something tangible like a ring of superconductor. So did London or other scientists run out and test the prediction?

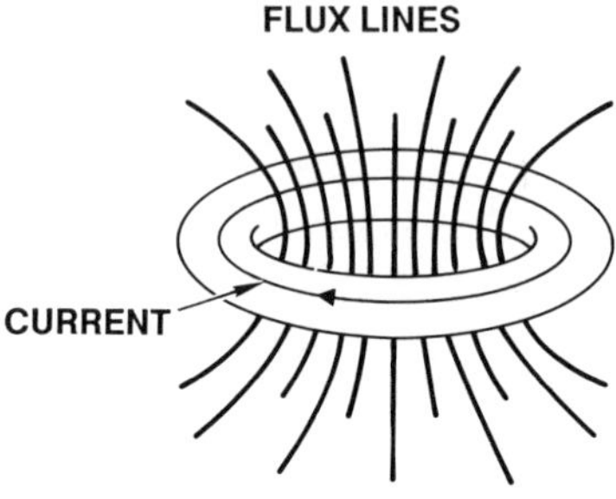

Magnetic flux threading a superconducting loop.

Not for about 11 years. For one thing, the prediction was hidden away in a footnote in one of London's books. More important, the flux quantum itself is extraordinarily small and in 1950, neither London nor anyone else knew how to make a small enough ring or to produce and measure small enough magnetic fields to see the effect. In the 1960's, however, these obstacles were finally overcome.

As we said, there are two characteristic lengths for superconductors that play an important role in our modern understanding of the phenomenon. We identified the first characteristic length as the magnetic penetration depth, developed in the London theory. Superconductors have a second important length called the *coherence length*, which was first proposed by Brian Pippard at Cambridge University.

In the early 1950s, a group at Cambridge researched the magnetic penetration depth in alloy superconductors. They found that the penetration depth strongly depended upon the nature of the alloy; in metallurgist's jargon, the "dirtier" the alloy was—the further from the pure metal form—the deeper the magnetic field would penetrate. The London theory could not account for these results.

In 1953, Pippard introduced a new formulation for the expression relating supercurrents and magnetic field. This formulation—known as nonlocal electrodynamics—contains within it a parameter known as the coherence length. Electrodynamics theory becomes a nonlocal theory when the electrical current at one spot must be calculated from the strength of the magnetic field (or more precisely, a related quantity called the vector potential) over an extended region rather than just at one spot. The size of the region is set by the coherence length, which in some sense measures the extent of the superconductivity. The consequence of this idea is that the density of superconducting electrons—or the two-fluid model's degree of superconductivity—

cannot appreciably change over an arbitrarily short distance in the superconductor, but only over a characteristic distance—the coherence length. Not only does this length play an important part in the successful Bardeen-Cooper-Schrieffer theory of superconductivity, but Pippard's equation of nonlocal electrodynamics itself appears virtually intact in the modern theory.

In the meantime, scientists became aware of other quantum aspects of superconductivity as well. The idea of an energy gap—which we discussed previously—began to appear in a variety of speculative theories of superconductivity in the 1940s. It was actually first proposed in 1935 by the Londons, who reasoned that what they called the "rigidity" of the superconducting wavefunction results from higher energy states being inaccessible. Good experimental evidence for the gap was not forthcoming until the 1950s when careful measurements of specific heat, thermal conductivity (a material's ability to conduct heat from one place to another), and the attenuation of ultrasonic vibrations in superconductors all indicated the existence of an energy gap. The most clear-cut evidence came from the far-infrared absorption measurements of Rolfe Glover and Michael Tinkham. We discussed these experiments before, in which a superconductor absorbs infrared light only when its frequency surpasses the gap energy.

The year 1950 was important in the history of superconductivity for a number of reasons. In that year, Vitaly Ginzburg and Lev Landau developed yet another phenomenological theory that, unlike its predecessors, remains a powerful tool for understanding many aspects of superconductivity to this day. The Ginzburg-Landau theory was developed as an alternative to the London theory of electrodynamics. However, in contrast with the essentially classical London theory, the Ginzburg-Landau theory makes use of quantum mechanics to predict the electrodynamic behavior of superconductors. It expressly makes use of a superconducting wavefunction whose attributes determine the behavior of the superconductor. In the language of the theory, this wavefunction is called an "order parameter" in the sense that it measures the extent to which the long-ranged order of the superconducting state exists. The Ginzburg-Landau order parameter is essentially equivalent to the London density of superconducting electrons or the Gorter-Casimir degree of superconductivity.

The details of the theory are overly mathematical for our purposes. What is important about the Ginzburg-Landau theory is that it remains the most successful theoretical tool for predicting the be-

Lev Landau. (Courtesy AIP Niels Bohr Library)

havior of superconductors whose properties vary over their length, of superconductors in contact with normal metals or other superconductors, and of a number of other systems in which spatial aspects play a key role. The theory was given only moderate respect for years until theorists showed that it could be derived from the BCS theory. Since then, it has become an accepted part of our theoretical understanding of superconductivity.

We have now covered the first 40 years of the development of the science of superconductivity. During that period, nearly all of the important properties of superconductors had been discovered. Superconductors exhibited zero resistance, the Meissner effect, a discontinuous specific heat at the transition, and a number of other effects. What was still needed was a theoretical explanation for how it all comes about.

The theoretical physicists of the 1950s had far more to go on than did their earlier counterparts. They knew that some kind of profound change was taking place in the electrons of a superconductor that results in a state with long-range ordering and a gap in the energy spectrum, and that this change was associated with a thermodynamic phase transition of the second order. At this point the stage was set for the development of the modern theory of superconductivity in 1957. All that was needed was one more key experiment to put it all together. We discuss that experiment next.

CHAPTER 5

Superconductivity Comes in Pairs

We now explore the question of exactly what happens inside a metal that turns it into a superconductor. Oddly enough, even an engineer who works with superconductors doesn't really need to know what makes superconductivity happen. If we understand what superconductors can do, then we are ready to start using them. Nonetheless, the microscopic theory of superconductivity developed by John Bardeen, Leon Cooper, and Robert Schrieffer in 1957 ranks as one of the great achievements of modern science and is well worth exploring.

What does a microscopic theory of superconductivity have to accomplish? To use an unflattering comparison, we might liken superconductivity to a disease contracted by a metal. For the past few chapters, we have been enumerating the symptoms of that disease: zero resistance, the Meissner effect, and so on. We even know that it takes cold to bring on these symptoms. A microscopic theory provides an explanation for how these symptoms come about on the atomic or electron level. Just as medical researchers want to know how a virus attacks body cells to cause a fever or inflammation, physicists want to know what makes the electrons in a superconductor behave in such an unusual fashion. The Bardeen-Cooper-Schrieffer (BCS) theory provides just this sort of explanation for the symptoms of superconductivity.

The BCS theory in fact proposes a microscopic mechanism that takes place within a superconductor that leads to all the diverse phenomena we have been talking about to this point. Although a complete accounting of the theory requires sophisticated mathematical formalism, its most profound physical insights can be explored on a more accessible level. We pursue that course in this chapter.

Creating a theory of superconductivity can be likened to baking a cake: we need the proper ingredients and we need a recipe. By 1950, a total of 40 years of research had provided the necessary ingredients for the theory but the recipe—how to put them all together—was still a mystery. The problem to be solved was to identify a microscopic mechanism—some kind of special behavior on the part of the ions and electrons that make up a metal—that leads to superconductivity. The superconducting energy gap and the evidence of long-range electronic order both pointed to the same idea: the electrons in a superconductor are somehow bound together. They want to act in unison. The question in physicists' minds in 1950 was, What brings them together?

The Role of the Lattice

Getting electrons together requires a special mechanism because they are negatively charged particles. By their very nature they repel one another. Physicists call this interaction the *Coulomb force*, which is repulsive between like charges. By the 1950s, sophisticated theories of the electronic properties of metals showed that such forces are tempered in the confines of a crystalline lattice; the influence of the positively charged ions in the lattice acts as a screen against the Coulomb forces between electrons. Nonetheless, even within a metal, electrons repel one another.

Thus the theory of superconductivity must furnish a mechanism by which electrons can cooperate with each other when electrical forces drive them apart. Could some unique physical alignment of atoms in superconductors do the job? Probably not. There are simply too many diverse and dissimilar metals, alloys, and compounds that exhibit superconductivity for such an explanation to make sense.

The right mechanism was proposed by the theorist Herbert Frohlich in 1950. He suggested that the ions in a superconductor— the crystalline lattice itself—must participate in the interaction of the electrons. In other words, superconductivity doesn't come about from a strictly electron–electron interaction, but rather from an electron–phonon interaction.

"Electron–phonon interaction" should ring a bell. We referred to it when we explained resistance in ordinary metals. Phonons are a fancy quantum-mechanical name for the natural vibrations of the crystal lattice in a solid. Phonons play the same role for ionic motion that photons play for ordinary light waves. We noted earlier that the disruption of the perfect crystal lattice caused by moving ions causes ordinary resistance in metals. Raise the temperature and the ions increase their motion. This increased motion presents more obstacles for passing electrons and the resistance goes up. If we describe the process in particle terms, we say that the electrons interact with phonons, so that the electron–phonon interaction causes resistance.

But now we are saying that the electron–phonon interaction causes superconductivity as well. There seems to be something self-contradictory about the same effect causing both resistance and the loss of resistance in metals. We will see that the fact that superconductors are always poor ordinary conductors and the best ordinary conductors do not become superconductors provides evidence for the role of the lattice in superconductivity. However, only a very compelling experimental result in 1950 convinced skeptics that the electron–phonon interaction was indeed the mechanism responsible for superconductivity.

Frohlich pointed out that the experiment to do was to look for an isotope effect in superconductors. Isotopes are alternate versions of a given chemical element that differ from one another in the number of neutrons in their atomic nucleus. The number of protons in the nucleus uniquely identifies a chemical element, so that changing the number of neutrons only results in a different atomic mass. Sometimes certain isotopes have very different properties from their lighter or heavier counterparts; for example, carbon-14, an isotope of carbon with two extra neutrons, happens to be radioactive. In most cases, however, the isotopes of an element are electrically and chemically indistinguishable from one another apart from their atomic mass.

From the standpoint of the ionic motion in solids, isotopes do make a difference. Heavier ions vibrate more slowly than their lighter counterparts. Scientists can calculate the frequency of an ion's vibratory motion using the same methods a freshman physics student would use to calculate the vibration frequency of a block hanging from a spring. The result gives a well-known formula expressing the relationship between the lattice vibration frequency and the mass of the ions.

If superconductivity results from electrons interacting with phonons—lattice vibrations—then changing the frequency of the vibrations should change the superconductivity. The concept is simple and, in principle, so is the experiment. When we measure the superconducting transition in ordinary tin, we find that it occurs at about 3.7 K. If we replace the tin with a heavier isotope (for example, tin_{122}, which has an atomic weight of 122 rather than the 118 of ordinary tin), the superconducting transition shifts slightly down in temperature. Change the way the ions vibrate and we change the superconducting transition. If we make such measurements for several different isotopes (and tin has nine to chose from), we find that changing the atomic mass alters the transition temperature in a very particular way. The precise way in which the temperature changes with isotope mass—the same relationship predicted by the freshman physics formula—is like the fingerprint of the electron–phonon interaction. If the superconducting transition temperature depends on isotope mass exactly the way the lattice vibration frequency does, then the two are intimately related.

Doing the isotope effect experiment required some brand new technology in 1950. What was needed was a way to separate out one isotope of a metal from another when most of their properties were indistinguishable. In the 1940s, nuclear physicists developed machines called *mass spectrographs* that could sort elements by their masses. Thus it was possible to separate isotopes in sufficient quantities to do superconducting isotope-effect experiments. The superconductors used then were the old standbys: mercury, lead, and tin. Many groups got into the act including the fiercely competitive laboratories of Kurt Mendelssohn at Oxford and David Shoenberg at Cambridge, as well as American groups at Rutgers University and the National Bureau of Standards. Emanuel Maxwell of the Bureau of Standards made the first measurements, but all the groups quickly succeeded in demonstrating the superconducting isotope effect. Thus, the experimentalists had obtained conclusive evidence that superconductivity was linked to the electron–phonon interaction.

How far had the theorists gotten by that point? Had they exposed the inner workings of superconductivity? Unfortunately not. Identifying the electron–phonon interaction as the mechanism responsible for an attraction between the electrons in a superconductor does not explain how such an attraction leads to superconductivity. It would take 7 more years to accomplish that task.

We now ask just how the electron–phonon interaction provides an attractive force between electrons. The complete answer lies in the mathematics of quantum mechanics. However, a simple physical picture illustrates how phonons bring about an electron–electron attraction. Anyone who has had the misfortune of sharing an old mattress with another person has firsthand experience with such a secondhand interaction.

When two people climb onto a sagging mattress, they create a depression in the gulf between them and they tend to fall into it. The mattress brings them together. More or less the same thing can happen to the electrons in a lattice. A passing electron attracts nearby ions because of their opposite charge. The electron thereby distorts the orderly arrangement of the lattice. Earlier we compared the ions in a lattice to a bunch of balls connected by springs. If we tug on one of the balls, the whole contraption will start oscillating. The distortion of the lattice caused by a passing electron is analogous to such a tug. Hence we anticipate that the lattice will vibrate, or as quantum mechanics puts it, we expect there to be phonons. Now comes the two-on-a-mattress analog. If another electron comes along at the right time, it encounters some ions shifted out of their normal positions in

A sagging mattress draws two sleepers together.

the vicinity of the first electron's path. In electrical terms, we say that the lattice has been slightly polarized. This extra concentration of positive charge tends to attract the new electron. The excess positive charge is like the depression in the mattress; the electron, like the second sleeper, tends to fall into it.

In this way we have a somewhat roundabout mechanism for getting two electrons to attract one another with a little help from the lattice. Physicists, in describing this process, say that the electrons are *exchanging momentum* with the help of phonons. What does that mean?

Momentum measures mass in motion. Apart from its vernacular meaning, momentum plays a central role in kinematics, the physics of motion. A fundamental law of kinematics states that when two objects collide, momentum is conserved. This means that if we add together the momenta for the two objects to find the total momentum for the pair, it will be the same before and after the collision. The momentum of each individual object can change but the sum of the two always gives the same answer. As momentum includes both the mass of an object and its velocity, a large mass moving slowly can have the same momentum as a small mass moving rapidly. Thus when a large truck crashes into a small car, the car lurches violently while the truck barely slows. The truck has transferred some of its momentum to the car, which, being lighter, gains more velocity than the heavier truck loses.

Physicists find it both useful and convenient to analyze the interaction between an electron and the lattice as a collision between two particles: an electron and a phonon. Such a description proves useful because collisions between electrons and phonons obey the same momentum laws as collisions between cars and trucks. The idea of two electrons exchanging momentum with the help of a third particle—in this case, a phonon—can therefore be understood in the light of what we know about more familiar objects. Picture two skaters facing one another on the ice. Suppose that one is holding a bowling ball (for some odd reason) and tosses the ball to the other skater who catches it. What happens? The two skaters move off in opposite directions. Why? The total momentum of the two skaters and the ball was zero to begin with; all three were stationary. Once the first skater tossed the ball, he gave it some momentum in the forward direction. That momentum was then transferred to the second skater when he caught the ball. The skater and ball move off together in the direction the ball was traveling, although at a lower speed than the ball alone (like the truck in the car–truck collision). Meanwhile the first skater recoils in the opposite direction so as to balance the books in total momentum. There was none to start with; there still must be none when we add up the total for the two skaters plus the ball.

This simple example demonstrates the principle of conservation of momentum, but for our purposes it demonstrates something else as well. The two skaters have repelled one another with the assistance of a third body: the ball. They have "exchanged momentum" and thereby changed their motion. They now move away from each other. Apart from the sign of the interaction, attraction versus repulsion, the electrons in a superconductor undergo just this sort of three-body momentum exchange.

Before we consider how this electron–phonon business relates to superconductors, we will briefly address the issue that a strong electron–phonon interaction appears to be favorable both to superconductivity and to high resistance. Have we encountered a paradox?

On the contrary. This state of affairs explains why the best ordinary conductors—silver, gold, and copper—do not become superconductors. Their very low resistance results from a weak electron–phonon interaction; the electrons in silver, gold, and copper are less likely to interact with ions than the electrons in other metals. However, a weak electron–phonon interaction means that superconductivity is unlikely to occur. The flip side of this argument holds as well. The best superconductors tend to be rather poor ordinary conductors because they have strong electron–phonon interactions.

The kind of lattice vibrations that cause resistance—that is, the phonons which are responsible—are called *thermal phonons* because they are stimulated by temperature. They occur at random and impart random momentum to passing electrons. The phonons that take part in superconductivity, on the other hand, do not owe their existence to thermal agitation of the lattice. The distortions of the lattice we are talking about are short-lived phenomena that come into being specifically to enable the attractive electron–electron interaction to take place. Nevertheless, both thermal phonons and the so-called virtual phonons that bring about superconductivity are just vibrations of the ions in the metal, differing in how and when they occur.

We have solved a key problem in the theory of superconductivity, namely how to attract electrons to one another. We have de-

A passing electron causes a lattice distortion leading to electron–electron attraction.

scribed a clever scheme that allows some ordinarily standoffish electrons to take a liking to one another by attracting some ions to the first electron and afterward having the second electron be attracted to the ions. If we focus our attention on the electrons, it looks like they are simply attracted to each other. Perhaps that much seems clear. However, the attractive interaction constitutes just the first step in explaining how metals come to be superconductors. The theorists needed additional insights to fill in the missing steps between such attracted electrons and superconductivity.

One theoretical physicist who wrestled with this problem was John Bardeen. Arguably no other physicist in this century has had a bigger impact on the development of technology than Bardeen. The only two-time recipient of the Nobel Prize in physics, he helped invent the transistor in 1947, thus initiating the modern electronic revolution. In later years he was responsible for key concepts in the development of the photocopy machine. In 1950, however, he turned his attention to the theory of superconductivity. At that time, he constructed a theory relating superconductivity to the electron–phonon interaction, duplicating Frohlich's results using a different theoretical technique. He then set out to fill in the rest of the theoretical picture of superconductivity.

In 1955, he pursued this research at the University of Illinois in collaboration with Robert Schrieffer, a graduate student. After awhile, Bardeen decided that a fresh outlook on the problem was needed and he looked for someone skilled in some of the new high-powered theoretical techniques being developed by particle physicists. He located a recent Columbia Ph.D by the name of Leon Cooper at the Institute for Advanced Study in Princeton.

Cooper Pairs

When we talked about thermodynamics, we said that physical systems always seek the lowest energy state they can occupy. Superconductivity occurs because the superconducting state has lower energy than the normal state. The theory of superconductivity had to provide a mechanism for this energy lowering. Leon Cooper took the next significant step in developing the theory of superconductivity in 1956 by calculating what happens when two electrons are added to a metal in the presence of an attractive interaction of the sort we have been discussing. The question is whether a lower energy-bound electron state will result. We follow the basic reasoning of Cooper's work.

If we add two electrons to a metal, they will have an energy greater than the Fermi energy, the highest occupied energy level in the metal. In terms of our familiar container full of marbles, two new marbles must sit above the top of the pile. Therefore, the total energy of the electron population—usually called the "Fermi sea"—increases with the presence of the two additional electrons.

If, on the other hand, the two electrons were somehow bound together, like the atoms in a molecule, for example, then they would *lower* the energy of the system. How do we make sense of this? Particles that are bound together must be held by some kind of a force. To unbind the particles, we must add energy in order to overcome that force and the amount of energy we add is called the binding energy. For example, the water molecule consists of two hydrogen atoms and an oxygen atom that are bound together. It takes energy to tear the molecule apart into its constituent atoms: the binding energy. Because it takes additional energy to break apart water, the water molecule must be in a lower energy state than the individual oxygen and hydrogen atoms. On the other hand, when hydrogen and oxygen combine to form water, an amount of energy equal to the binding energy will be released. Devices called *fuel cells* aboard spacecraft combine hydrogen and oxygen to form water and thereby generate energy by just this principle.

According to classical physics, any two particles, no matter how weakly attracted, can form a bound state, lower in energy. In contrast, tightly bound quantum mechanical particles have large average velocities and thus large energies of motion. Unless the attractive binding energy is sufficient to overcome this kinetic energy, no bound state will be formed. This was the puzzling aspect of superconductors. The attractive interaction brought about by phonons was far too weak to produce a bound state.

What Leon Cooper was able to show was that in the presence of an attractive interaction, no matter how weak, two electrons added to the Fermi sea will form a bound pair. Even though the kinetic energy of the added pair is higher, the overall energy of the system is lowered by the formation of the bound pair. This implies that the Fermi sea is unstable against the formation of bound pairs—it wants to form them—because the energy of the system could be lowered by taking two electrons of a given energy from the top of the sea and putting into a bound state formed from energies above the sea. The presence of the other electrons of the Fermi sea allows this to happen; two isolated electrons in the presence of a weak attractive interaction could not form a bound state. Scientists call such pair formation a *many-body effect*.

Such pairs of electrons in a superconductor turn out to be very weakly bound indeed, as the amount of energy required to break up a pair is only a few thousandths of an electron volt (the usual units for electron energy). In comparison, the Fermi energy of the electrons in a metal is generally 5 to 10 electron volts. So the pair-binding energy constitutes only about a thousandth of the electron energy. We might liken this to hooking two cars together with a paper clip; such pairs must break apart very easily.

How can pairs that come apart so readily lead to any kind of mass behavior by the electrons in a superconductor? The answer is that under the proper conditions, the pairs continually recombine in order to maintain a state of lowered energy. What matters is that, on the average, electrons spend their time in paired states. Later, we examine the conditions for formation of pairs.

We can most easily visualize the electron–phonon interaction in terms of passing electrons pulling nearby ions out of their normal positions, but for the purposes of theoretical analysis, scientists use a different picture. The interaction that attracts the electrons to one another can be viewed as a scattering event involving two electrons and a phonon. The same sort of scattering occurs when two billiard balls collide; they go off in new directions. Here is how we describe the electron–phonon interaction as a scattering event. The first electron collides with a phonon (it interacts with the lattice), transfers some of its momentum to the phonon (it distorts the lattice), and gets scattered into a new trajectory. The second electron then collides with the same phonon (it runs into the distorted part of the lattice), picks up the momentum originally given up by the first electron, and also gets scattered into a new trajectory. We are reminded of our skaters with the bowling ball except that this time the net interaction is attractive. The important thing about this three-body interaction is that it lowers the energy of the electron system. Such an interaction tends to happen because nature always seeks the lowest energy state.

The ground rules for this little game of pool require that the total momentum be conserved in the scattering process but do not demand energy conservation for the incoming and outgoing electrons. As long as the energy difference between the electrons is no more than the binding energy of the pair, such scattering events will lower the energy of the system. Within these restrictions, the trick is to maximize the number of scattering events in order to maximize the energy lowering of the system.

Cooper showed that the energy lowering from the formation of a bound state will overcome the additional kinetic energy of the electrons when the electrons of a pair have equal and opposite momentum. What does that mean? An example of two objects with equal and opposite momentum are a half-ton truck headed north at 55 mph and a half-ton truck headed south at 55 mph. Add up these two momenta and we get zero; the momenta are equal and opposite. The largest number of energy-lowering scattering processes can occur between electrons of equal and opposite momentum. Since the publication of Cooper's 1956 paper, such bound pairs of electrons of equal and opposite momenta have been known as Cooper pairs.

Cooper pairs are the second key element in the theory of superconductivity. In his paper Cooper had considered the artificial case of two new electrons added to a metal. The task awaiting BCS was to treat the case of the 10^{23} electrons per cubic centimeter already present in a real metal. When John Bardeen returned to Illinois from the Nobel Prize ceremonies in Stockholm in late 1956, he was ready to tackle the problem.

The BCS Theory

The interactions of large numbers of electrons presents an intimidating subject for theoretical analysis; such *many-body effects* represent some of the most difficult problems in physics. In order to develop a theory that could predict the behavior of the multitude of electrons in a superconductor, Bardeen, Cooper, and Schrieffer made a simplifying assumption. Rather than attempting to analyze the effects of interactions between three, four, or more electrons at a time, they assumed that only pair interactions matter. So, in considering the behavior of a pair, the presence of all the other electrons merely limits the states into which an interacting pair can scatter (because the electrons still obey the Pauli Exclusion Principle). Relegating the other electrons to a background role results in what solid-state physicists call a *mean field theory*.

By making this mean-field assumption, the BCS theory treats all the electrons in a metal at once. Bardeen, Cooper, and Schrieffer proceeded to combine the concepts of the attractive electron-phonon interaction and the Cooper pair with an appropriate mathematical structure that produced an accurate description of the superconducting state. They were able to isolate the important although weak interactions that lead to superconductivity in a rigorous theoretical model. The BCS model has proven so successful in describing certain many-body systems that it was later successfully applied to the seemingly unrelated problem of the binding of the protons and neutrons in the atomic nucleus. We now examine the BCS theory in more detail.

As we have said, an attractive interaction results in the Fermi sea of electrons being unstable against the formation of bound pairs of electrons formed from states above the Fermi level. Therefore, it is necessary to find a new ground state (lowest energy state) of the system in which many pairs of electrons of opposite momenta are in higher energy states than the maximum of the normal Fermi sea.

To describe this superconducting ground state, the BCS theory has to provide for the formation of more than one Cooper pair in a metal.* We focus our attention on a pair of electrons sitting at the top

---

*To simplify the description, we will consider the metal to be at absolute zero temperature so that if we measured the energy of the electrons, none would have more than the fermi energy (the highest filled energy level). At higher temperatures, some of the electrons would be above the fermi level because they would have additional energy from thermal excitation. In terms of our familiar marbles analogy, at absolute zero all the marbles would be sitting in their neat rows; at finite temperature, the marbles at the top would be jiggling up and down, upsetting the stack.

Nobel Prize winners John Bardeen, Leon Cooper, and Robert Schrieffer. (Courtesy AIP Niels Bohr Library)

of the Fermi sea, just below the Fermi energy level. These are the most accessible electrons to use for forming pairs. Can we scatter two of them off of a phonon and form a Cooper pair to lower their energy further? If the process results in a lower energy state, we can first promote two electrons from energy states below the Fermi level up to vacant states above the Fermi level, and then let them interact to form a Cooper pair. We first raise the energy of these electrons because they have to scatter into new states (change their trajectory) when they interact. Only above the Fermi level will we find an appreciable number of vacant states to scatter into. Below the Fermi level, the electrons have nowhere to go.

Once we have a way to build pairs, we can continue the process and form even more. If forming one pair lowers the energy of the system, forming more pairs lowers it even further. We simply take pairs of electrons with equal and opposite momenta from below the

Fermi level and let them scatter off of phonons, lowering their energies in the process. We thereby form a new state made up of Cooper pairs. Can all the electrons in the metal participate in the process? That creates a problem. It would imply that the way to lower the energy of the electron system begins by raising all the electron energies above the Fermi-energy level, which seems preposterous.

We are saved from an absurd predicament by the standard requirement for quantum mechanical processes that scattering has to take place from full states into empty states. As more and more electrons form Cooper pairs, there become fewer and fewer empty states to scatter into and, as a result, the number of possible scattering processes decreases. There are fewer possible ways to rearrange the electrons. With fewer scattering processes, the amount of energy-lowering decreases and finally we reach the point at which the energy price we pay for raising electrons above the Fermi level is greater than the energy dividend we receive from forming a Cooper pair. At that point, we no longer profit by forming any more pairs. Thus we find an optimum number of Cooper pairs that will form in a superconductor.

The most important property of the BCS ground state is that the pairs are completely interchangable and indistinguishable from one another; the paired electrons can all be described by a single quantum-mechanical wavefunction. Pairs continually scatter from occupied pair states into unoccupied pair states; such scattering events occur so long as each pair has the same total momentum as every other pair. In quantum-mechanical terms, this BCS ground state is a *coherent mixture* of single-electron wavefunctions.

We have just followed an arduous descriptive path to what is known as the BCS superconducting ground state, so called because it has the lowest energy. The path that BCS took was one of mathematical complexity and theoretical sophistication, but one that leads to the same result: a new state of matter in which the electrons in a metal are paired up in a very particular way to produce a lower energy state in the metal. Shortly we will see how this electron-paired state results in superconductivity, but first we look at a few more features of the BCS ground state.

With some elementary quantum mechanics, we can calculate a nominal spatial extent of the Cooper pair, which corresponds to the coherence length proposed by Brian Pippard. It turns out to be many thousands of atomic lengths. In a typical metal, a three-dimensional volume of this size contains millions of electrons. Therefore the Cooper pairs must overlap in great multitudes. The overlapping of so many pairs has a number of consequences. We have already seen that the energy of the system will be the lowest if the greatest number of electrons participate in the pairing process. If all the pairs are alike, then nothing stands in the way of partners from one broken pair finding new mates from other broken pairs. The condition in which large numbers of pairs have the same constant total momentum and occupy the same physical space provides the greatest opportunity for the electrons to recombine.

The BCS theory allows us to calculate the probability for forming Cooper pairs from electrons of given energy levels in the Fermi sea and tells us what the remaining occupied states must look like. This calculation reveals an energy distribution of just the sort theorists were expecting for superconductors, complete with an energy gap. In the context of the theory, the gap corresponds directly to the binding energy of the Cooper pairs. It refers to the amount of energy that must be added to a superconductor to break up a pair and create individual electrons. This corresponds to adding energy to a water molecule in order to turn it back into oxygen and hydrogen. In a superconductor, the presence of a gap implies that we must break up a Cooper pair before we can give any energy to the individual electrons. If this were not so, and we were able to add an arbitrarily small amount of energy to a single electron near the Fermi level, then that electron would be excited into a higher energy state and could no longer participate in the formation of pairs. The binding energy for one pair would then be lost to the system. Instead, the minimum energy required to excite a single electron to a higher energy state equals the binding energy for a pair and we have a gap in the energy spectrum.

If we make the analogy of superconductivity to a disease, we have now succeeded in finding the virus that causes it. We don't yet know where all the symptoms come from, but having identified the bug that causes them, we can track the symptoms down. At this point, let us review and amplify what we have just explained as we try to form an understandable picture of the BCS superconducting state.

In a superconductor, an electron passes nearby ions and gets thrown off course in a process that transfers some of its momentum to the lattice. Such a momentum transfer has the physical consequence of distorting the lattice so that it locally concentrates some positive charge. A second electron feels the influence of this polarization of the ions and gets drawn toward the excess positive charge. Its course becomes altered so as to pick up the momentum imparted by the first electron. This two-step process results in a slight attraction between two electrons.

The BCS theory takes this process further and shows that such an attraction can overcome the natural repulsive force between electrons and produce a bound state known as a Cooper pair. Such pairs are bound so weakly that they are constantly breaking apart and recombining with new partners. Each pair actually occupies a large region of space, one that contains many other electrons that could participate in the pairing process. The system achieves the tightest binding—has the lowest energy—if the greatest number of electrons participate in the pairing process. The most advantageous way for this to happen results from all the pairs having the same zero momentum.

Once again we find that the principle of least energy determines the dynamics of a physical system. The lowest energy state of a superconductor—known as the BCS ground state—consists of Cooper pairs of electrons with equal and opposite momenta. The binding energy of these pairs shows up as a gap in the density of electronic states for the superconductor. In order to give energy to individual electrons, pairs of electrons must break up, a process requiring an energy equal to the gap value.

These three paragraphs sum up what we know about the BCS ground state. Now we need to consider what the consequences of such a state must be.

Explaining Superconductivity

The central point of the BCS theory is that the superconducting ground state constitutes a real state of matter, like liquid or solid. The atoms of a liquid freeze into the solid state when it becomes energetically favorable for them to adopt the ordered form of the crystal lattice. Similarly, the electrons in a superconductor "condense" into the BCS ground state when a lower energy state results from the binding of Cooper pairs. The condensation in this case does not refer to a condensation in physical space, like the condensing of a gas into a liquid, but rather to a condensation in momentum space. Momentum space is simply a mathematical construct used by physicists for describing the motion of the constituents of solid matter. In simple terms, a condensation in physical space means that particles cooperate with regard to where they are; a condensation in momentum space means that particles cooperate in how they move.

An analogy may help clarify the rather abstract concept of a momentum-space condensation. Compare the following two images: a teeming crowd of people in a busy airport terminal and a few thousand cadets marching in step at a military academy. The motion of electrons in an ordinary metal might be likened to the airport crowd working its way towards the gates. The condensed electron state in a superconductor would look more like the synchronized marchers at the academy. Synchronized motion typifies a momentum-space condensation.

This condensed state's most remarkable property is that in order to achieve the maximum lowering of the energy provided by the pairing process, the electron pairs have essentially forsaken all remnants of their individual identity. Because they all continually scatter in and out of a narrow band of energy states near the fermi energy, we cannot distinguish between them. They are like the interchangeable links in a moving bicycle chain. The theory of quantum mechanics describes electrons in terms of wavefunctions, a separate one for each electron. The BCS theory demands that all the paired electrons in a superconductor be regarded as belonging to the same quantum state described by the same identical wavefunction. Cooper pairs do not have to obey the Pauli exclusion principle that restricts individual electrons to separate quantum states. The pair state corresponds to the macroscopic quantum state predicted by Fritz London and sought by theorists since the 1930s.

The state so produced constitutes a highly ordered state. All the pairs behave in the same way; they have the same motion and same wavefunction. This order does not just extend over the size of a pair. The pairs overlap in such a multiple fashion that the ordering propagates throughout the lattice. The process can be likened to laying floor tiles in a large room. The orderly arrangement of the tiles at one end of the room propagates across the room clear to the other side. The tiles at one end of the room have dictated the arrangement of the tiles at the other end. In the same way, the ordering of the pairs in a superconductor propagates through the lattice.* This is what we mean by long-range order.

At last we can explore the consequences of the BCS ground state. In thermodynamic terms, the BCS ground state provides us with a state with lower energy than the normal metal state, and so the metal prefers to be in that state. The lower energy state is a many-body quantum-mechanical state in which all the electron pairs are described by the same wavefunction. Surprisingly, that simple fact accounts for all the unique properties of superconductivity. Such properties are not the essence of the many-body state, they are a reaction to it. All the remarkable features of superconductors come about from the metal trying to remain in the reduced-energy BCS ground state. We examine how this principle applies to several key superconductor properties.

The most important case to analyze is the response of the BCS superconductor to an applied electric field. We recall what happens in a normal metal. When we apply an electric field to a normal metal (which we do by applying a voltage across the sample), the electrons accelerate and current begins to flow. The electrons then collide with imperfections in the lattice as well as thermal phonons and they lose their energy to these resistive processes. We eventually reach a trade-off situation in which we continue to supply energy from the applied voltage but that energy in turn gets consumed in heating up the metal. We call this process *Joule heating*. We need voltage to produce current and the metal obeys Ohm's Law. If we turn off the energy source, the current dies out.

We try the same thing with the BCS superconductor. What happens? The electrons have two options: they can individually bump into impurities, phonons, or the like, and thereby give up the energy advantage gained by pairing up. Or they can continue to move in unison, this time with a common nonzero momentum of the pairs set by the external field. As you may guess, they take the second option, although it is not a capricious choice. The object of the game is still to minimize the energy of the system. The lowest energy state of the system remains the one in which the motion of each pair gets locked

---

*Keep in mind that this ordering is nothing like the ordering of ions in a lattice. The pairs do not arrange themselves in specific positions, they are ordered in momentum space.

into the motion of all the rest. Thus stragglers are not allowed to go off on their own and crash into various obstacles in the lattice. If one member of a pair gets bumped along the way, its partner responds in such a way as to maintain the original bearing of the pair. Cooperation pays off. Any camp counselor will tell you that the best way to get a pack of children across a crowded shopping center is to have them link hands and move as a group. Cooper pairs use the same basic trick to traverse a lattice unscathed.

In this way the BCS-paired ground state encompasses within it the zero-resistance property of superconductors. The argument boils down to energy considerations. If remaining paired and moving in unison results in a lower energy state, then the electrons will sustain that state. Of course accelerating the electrons adds some energy of motion. If too much energy is added, then the pairs must break apart. We call the amount of electric field needed to break up pairs the *depairing field*; in practice, breaking pairs requires currents so large that the superconductor has long since surpassed the critical current associated with its critical magnetic field.

Such arguments about zero resistance in superconductors are really arguments about the energy advantages of keeping the electrons linked together. They don't preclude the possibility of *all* the Cooper pairs scattering off impurities or the like at once. Of course, even a tiny mote of superconductor contains trillions of pairs. But if they were all to simultaneously scatter, there would be a loss of energy to the lattice and the current in the superconductor would decay. Theoreticians are quick to point out that the chances of such a wholesale scattering event taking place within the lifetime of our universe are extraordinarily slim. The resistance of a BCS superconductor can be safely considered to be zero.

The BCS theory provides a natural explanation for the specific heat behavior of superconductors as well. Experiments show that at the onset of superconductivity, it suddenly takes extra energy to heat up a superconductor. We now can see that if we add energy to a superconductor, it does not immediately raise the temperature of the material. Energy first must be used to break up pairs. Thus at the temperature at which pairs first appear—the critical temperature— the amount of energy required to raise the superconductor's temperature abruptly increases. We notice that this argument really boils down to a restatement of our earlier explanation of the energy gap in a BCS superconductor.

As far as other superconducting phenomena are concerned, the BCS theory provides a firm underpinning to the theories that already were successful in the past. The London theory, Pippard's nonlocal electrodynamics, and the various two-fluid models that we talked about earlier are all phenomenological theories that account for the behavior of superconductors. BCS did not prove them wrong; it enhanced them by explaining what was going on inside the superconductor to cause the phenomena. In particular, there were no longer mysterious superconducting electrons to contend with. Zero-resistance currents are sustained by Cooper pairs and the strength of superconductivity really measures the strength of the pairing.

The strength of superconductivity, seen in the way in which various superconducting parameters vary with temperature, comes naturally out of the BCS theory. The most important prediction of the theory regarding a temperature-dependent property involves the energy gap and its relation to the superconducting critical temperature. The reason that the gap varies with temperature at all can be easily understood. We start with a superconductor at absolute zero. Its ions are nearly motionless at this lowest temperature in nature (quantum mechanics still requires some movement known as zero-point motion). As we warm the material up, the thermal agitation of the ions intensifies. At first, this increased ionic motion has little effect. A few less pairs might be around, but the superconductivity remains essentially undisturbed. But with rising temperature, more and more pairs break up and the individual electrons, with their additional thermal energy, are no longer available for pairing. With fewer possible scattering events, the energy lowering brought about by pairing diminishes. Since the gap measures this binding energy, it too decreases with rising temperature. The critical temperature defines the temperature range that matters. Far below the critical temperature, these effects are negligible; near it, they are extreme. The critical temperature itself marks the point at which all energy advantage in pairing disappears. At that temperature the energy gap becomes zero.

The BCS theory goes further with this argument and predicts a direct relationship between the energy gap and the critical temperature of a superconductor. The theory tells us what the energy gap of a superconductor will be if we know its critical temperature. According to the theory, the two are proportional and have a predicted ratio. Most superconductors actually exhibit the predicted ratio between

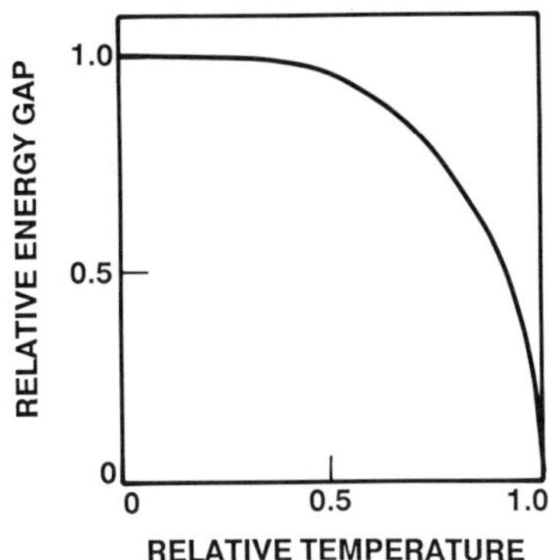

The BCS energy gap decreases with rising temperature.

the critical temperature and the energy gap and are known as BCS-type or "weakly coupled" superconductors. Certain other superconductors—for example, lead—have somewhat larger energy gaps than the theory would predict and are called strong-coupled superconductors. These details have grown out of later theoretical developments in superconductivity.

The development of the BCS theory provided the much-sought-after microscopic model of superconductivity. John Bardeen, Leon Cooper, and Robert Schrieffer were awarded the Nobel Prize for their work on the theory. But the BCS theory was not the last major theoretical work to be done in the field, nor were all previous efforts discarded. As we said earlier, a number of problems in superconductivity can still only be treated by the theory developed by Vitaly Ginzburg and Lev Landau. However, that theory attained a whole new status when it was shown that it could be justified by the principles of the BCS theory. Another question that BCS could not answer was how to predict which substances become superconductors and at what temperature. We return to this question later on.

Beyond BCS

As far as later developments in the theory of superconductivity are concerned, several important Soviet physicists expanded upon the BCS theory in order to more realistically treat the interactions and excitations in the system. Nikolai Bogoliubov devised a powerful formalism for describing the excited single electron (or more properly, quasiparticle) states in superconductors. George Eliashberg produced a sophisticated treatment of the electron–phonon interaction and its role in the pairing process. These results lead to the distinction between weak- and strong-coupled superconductors that we mentioned before and that was borne out by the electron tunneling experiments of the early 1960s. The strong-coupling theory was essential in explaining discrepancies between the BCS predictions for the tunneling density of states and what was observed in real lead tunnel junctions.

The other major theoretical development in superconductivity came in 1962 with Brian Josephson's prediction of pair tunneling between weakly linked superconductors. Josephson was a Cambridge graduate student who received the suggestion from Brian Pippard that the problem of Cooper pair tunneling in a superconducting tunnel junction would be worth studying.

Tunnel junctions had first been studied by Ivar Giaever at General Electric Laboratories in 1960. A theory worked out by Giaever and his colleague Kenneth Megerle described the tunneling process in terms of the quantum mechanical process that we described earlier. According to the theory, the chances of a particular electron making it through a tunneling barrier to get to the other side were slim but nonzero. Tunneling of single electrons should happen often enough to be seen in real experiments. The conventional wisdom on the subject of pair tunneling, on the other hand, was that the odds of two electrons making it across in tandem were far worse. Elementary probability theory tells us to multiply the odds for two events when we want to know the chance for both occurring. For example, the odds for a coin flip coming up heads are one in two. The odds for two flips in a row coming up heads are only one in four. For this reason, it was thought that pair tunneling was so much more unlikely than single electron tunneling that it would never be observed.

In a brief paper, Josephson showed that the problem of pair tunneling was not like coin flips at all. A better analogy would be having a red marble and a blue marble in a bag and asking the odds for pulling out a red, then a blue. The answer, of course, is one in two. There is a fifty-fifty chance for the first marble to be the red one, but if it is, the second one is guaranteed to be blue. Similarly, the second electron in a Cooper pair doesn't face the usual odds against tunneling. The energy advantage to remaining paired carries over across the tunneling barrier and it turns out to be no more unlikely for a pair to tunnel than a single electron.

Nobel Prize winner Ivar Giaever. (Courtesy AIP Meggers Gallery of Nobel Laureates)

By 1963, the first Josephson junctions had been made and the theoretical prediction was verified: pairs could tunnel across a barrier as easily as single electrons. Of course, as we now know, when pairs carry current, there is no resistance. Thus there is a zero-voltage current in a Josephson junction. Josephson also predicted an additional effect in these junctions: the so-called ac Josephson effect. When a voltage is applied across a thin barrier between two superconductors, an alternating current flows whose frequency is set by the size of the voltage. This too was verified by experiment. In recognition for all these significant theoretical and experimental developments in superconducting tunnel junction, both Brian Josephson and Ivar Giaever were awarded the 1974 Nobel Prize in physics.

We return to the subject of Josephson junctions in the next chapter. At that point we also explain the third major theoretical develop-

Nobel Prize winner Brian Josephson. (Courtesy AIP Meggers Gallery of Nobel Laureates)

ment in superconductivity that took place between 1957 and 1962: the theory of Type II superconductors. We address both of these topics when we discuss the nature of superconducting materials and how they are made into practical forms.

CHAPTER 6

Practical Superconducting Materials

Bringing superconductors out of the laboratory and into the real world requires the skills of many disciplines. The scientific discoveries of solid-state and low-temperature physics must be developed into superconducting technology by material scientists, electrical engineers, vacuum scientists, and others. And the first step in creating a superconducting technology is the development of technologically useful superconductors.

At this point, we explore the topic of superconducting materials and how they have evolved to meet the needs of superconducting techology. We only talk about the materials that have been developed for practical use; we defer the topic of more exotic superconducting materials until later. With a few notable exceptions, applications of superconductivity require one of two forms of superconductor: thin films or wire. Each form has its own unique history, so we trace the development of both.

The history of superconducting materials goes back to the pioneering work of Kamerlingh Onnes. The first superconductors to be discovered—mercury, tin, and lead—were extensively studied in Leiden in the early decades of this century. Such materials soon proved to be of little practical value for the applications that scientists immediately envisioned for superconductivity. Onnes wanted to build immensely powerful magnets from superconducting wire, but the first superconductors could not remain superconducting in even moderately large magnetic fields. Superconductors like tin or lead work very well for superconducting electronics, on the other hand. However, the fundamental devices of that technology would not be discovered until the 1960s.

Thus, almost from the outset, researchers needed new and better superconductors. Two driving forces dominated the quest for new superconducting materials: finding ones that could withstand high magnetic fields and finding ones that operated at higher temperatures. The two goals coincided to the extent that superconductors can withstand the highest magnetic field when they are far below the temperature at which they become superconducting. When the critical temperature of a superconductor is high, we can easily cool it down to a small fraction of that temperature and thereby enhance the critical field—the magnetic field at which superconductivity disappears. But the coincidence of high-critical field and high-critical temperature goes further because the highest temperature superconductors tend to have the highest critical field. Thus, the real challenge boils down to finding new, higher temperature superconductors.

In addition, we must also realize that little of a useful nature can be done with a superconductor at or near its transition temperature—the temperature at which it becomes superconducting. We have made reference to the idea of the strength of superconductivity and how it changes with temperature. This is not merely a subject of academic interest, but it plays a crucial role in all superconducting applications. Virtually any superconducting property we wish to exploit—capacity to handle high current, ability to withstand high magnetic fields, or even the zero-resistance Josephson current—vanishes at the transition temperature and reaches maximum strength at the lowest temperatures. In fact, most of these superconducting properties change dramatically as we cool the superconductor from its transition temperature down to a little above half that temperature. As a rule of thumb, superconductors give their best performance at temperatures below about 60 percent of their critical temperature. Therefore, a 7 K superconductor like lead only becomes useful around 4 K, the temperature of liquid helium. At 6 degrees, for example, lead has very poor superconducting performance. So the need for higher temperature superconducting materials is stronger than it might first appear. Not surprisingly, the search for them has gone on for decades.

Finding New Superconductors

When scientists tested more and more materials for superconductivity, they uncovered a surprising fact: the phenomenon is not rare. In fact, over a quarter of all the natural elements—and more like half the metallic elements—are known to superconduct at low temperatures. Some of these are garden variety elements like tin and lead, while others are truly obscure elements like iridium and ruthenium. The temperature at which superconductivity occurs in the elements varies from as low as a hundredth of a degree Kelvin in tungsten to as high as more than 9 degrees in niobium.

Niobium has become the most important superconducting element for reasons of its physical durability and relatively high superconducting temperature, yet if it wasn't for its superconducting applications, few of us would have ever heard of this heavy grayish metal. Until quite recently, niobium was called columbium after Columbia (meaning America), because it was discovered in 1801 in a 100-year-old ore sample sent from America to England. Superconducting niobium retains its superconductivity in a higher magnetic field than any other pure element, but it is still not high enough for making powerful superconducting magnets. Therefore, while elemental niobium continues to be a mainstay of the superconducting electronics world, its main contribution to both the science and technology of superconductivity comes from the many niobium-based superconducting compounds.

One can easily exhaust the list of the 20-odd superconducting elements. In contrast, there is practically no limit to the number of intermetallic compounds and alloys—formed from two or more metal elements—that we can investigate. Scientists began to produce samples of superconducting compounds in the 1920s and 1930s. Today, a complete listing of superconducting intermetallic compounds would fill many pages of this book. There are literally thousands of superconducting compounds composed of several metallic elements, or in some cases, formed from combinations of metals and nonmetals. Many of the highest temperature superconductors discovered have been compounds containing niobium. Three years before his famous magnetic field experiment, Walther Meissner found superconductivity above 10 K in the compound niobium carbide. The first so-called high-temperature superconductor was probably niobium nitride, a 16 K superconductor first reported in 1941. Nevertheless, finding other such compounds proved to be a difficult task in the absence of theoretical principles to guide the search.

We have already explored the many successes of the BCS theory of superconductivity. In fairness, we have to mention its greatest shortcoming: it can't predict which materials turn out to be superconductors. In truth, no theory exists that can allow us to consider a new material and decide whether it becomes a superconductor at some temperature. This might seem surprising in the light of the acknowledged success of the BCS theory.

The theory explains how materials become superconductors but it does not predict which materials will do it. The BCS theory describes the electron–phonon interaction in only a very simplified way; what really goes on in a metal is much more complicated. We can't predict which materials have the necessary interactions for superconductivity because the only way we know how to test a material for the presence of these interactions is to see if it superconducts. In other words, we determine whether a substance can superconduct by looking for superconductivity in it. Unfortunately, we don't know how to test some high-temperature property of a substance and accurately predict when and whether it will become superconducting. No fundamental prohibition exists against such a test; we just don't know of one.

Nevertheless, looking for new superconductors does not have to be a blind search. Over the years, scientists have developed empirical rules that help locate the likeliest candidates for new superconducting materials. Most of these rules—and many of the new materials—were discovered by a scientist named Bernd Matthias. Empirical rules are rules discovered from experiment, not deduced from theory. What Matthias and co-workers like John Hulm did in the 1950s was to study large numbers of different materials and look for patterns in their behavior. At first there didn't seem to be any. Some compounds superconducted while others didn't, without any apparent rhyme or reason. But after 3 years of laborious tests, a pattern finally emerged.

They found that the number of valence electrons in the atoms of the material was important to superconductivity. According to the modern theory of the atom, electrons occupy shells that surround the

Bernd Matthias. (Courtesy AIP Niels Bohr Library)

nucleus. In most atoms, the inner shells are full—they hold as many electrons as they can—while the outermost shell is often only partly occupied. The electrons in the outermost shell—the ones that take part in the conduction of electricity—are called *valence electrons*. Matthias observed that the only substances that became superconductors were elements or compounds with specific average numbers of valence electrons per atom. We speak of averages because compounds contain different atoms that may have different numbers of valence electrons. The average number of valence electrons in the atoms making up the compound turns out to be the deciding factor. By comparing the superconducting transition temperatures of many different materials with the average number of valence electrons per atom of the material, Matthias found that materials with averages of near 5 and 7 electrons-per-atom were the highest temperature superconductors.

This bit of numerology provides an empirical formula for finding new superconductors; no one knows why it works. Nonetheless, when this rule was applied in conjunction with the observation that certain crystal structures are also favorable for superconductivity, the results were quite astounding. Matthias and his co-workers discovered many new superconductors by synthesizing new compounds according to these rules. We cite a specific example to illustrate how it was done.

The metal ruthenium has 8 valence electrons per atom and becomes a superconductor at half a degree above absolute zero. Molybdenum, a metal with 6 valence electrons, also superconducts below 1 degree Kelvin. However, an alloy containing equal parts of the two has an average of 7 valence electrons and it superconducts at 10.6 K. Producing the right average number of electrons per atom leads to a better superconductor. Matthias exploited this kind of behavior to the maximum. In some cases, he produced superconductors from the combination of an element that doesn't superconduct with another element that doesn't even conduct at all. A cobalt-silicon compound that superconducts demonstrates this kind of behavior.

Synthesizing compounds according to Matthias's rules became a major research activity in the 1960s and 1970s. The quest for the highest possible superconducting transition temperature soon centered around the niobium compounds and one particular crystal structure that crystallographers call the beta-tungsten or "A15" structure. This particular arrangement of atoms in a crystal appears in compounds that contain three atoms of one element and one of another. The elements are arranged in a cube such that one element occupies the corners and the center of the cube and the other forms linear chains in three dimensions.

This arrangement of atoms has proven to be very favorable for superconductivity, primarily because it provides a particularly high density of electronic states that can participate in the formation of Cooper pairs. The first superconductor of this sort scientists discovered was a vanadium-silicon compound, which they found to superconduct at 17 K in 1952. Over the next 20 years, scientists discovered more than 20 more superconductors in the A15 family.

Several of these compounds—especially those containing niobium—flirted with the 20-degree mark for superconductivity. This was something of a magic number because hydrogen liquefies at 20 K; a 20-degree superconductor would be the first that did not require

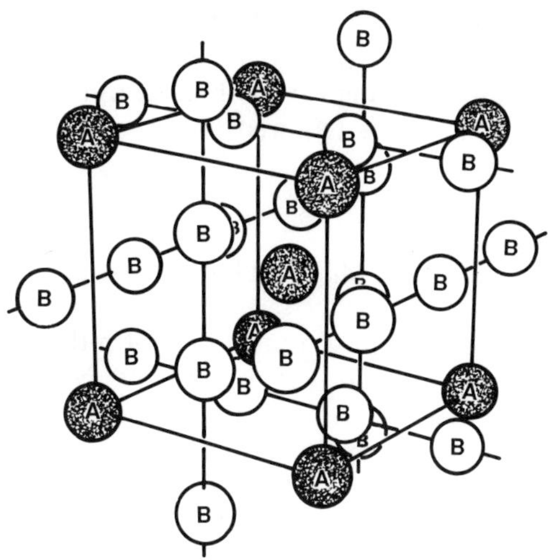

The A15 Crystal structure for the compound "A_3B."

liquid helium. As it turned out, breaking the 20-degree barrier required innovations in materials research. The ideal A15 crystal structure simply would not form in some of the most promising niobium compounds. For energy minimization reasons, the atoms would rearrange themselves into other structures instead. For this reason, researchers needed to produce metastable phases of the desired compounds. Such phases are formed when the elements are forced into the proper arrangement during a high-temperature crystallization process and then trapped into that arrangement by rapid cooling. The rapid cooling techniques prevent the atoms from relaxing into arrangements that are unfavorable for superconductivity. By such a technique, George Webb broke the 20 K barrier at RCA's research lab with his discovery of superconductivity in a niobium-gallium compound. John Gavaler was able to synthesize a niobium-germanium A15 compound in 1971, a material that held the superconducting temperature record for the next 15 years at 23 K . Despite the hope for even higher critical temperatures in other intermetallic compounds, it would take a breakthrough in an entirely different class of material to surpass the superconductivity of the niobium compounds. We will defer the story of this breakthrough until later.

The search for higher temperatures was not the only activity in the field of superconducting materials. The 1960s saw the true birth of applied superconductivity with the advent of two technologies: Josephson junctions and high-field, high-current superconducting magnets. Both technologies required important advances in superconducting materials.

Making Films

As we note later, Josephson junctions form the basis of a diversity of superconductive electronics applications. Naturally, before we can think of using Josephson junctions, we need to be able to make them. Josephson junctions and superconductive electronics in general require a thin film technology. We may not all be familiar with thin films but they nonetheless play a growing role in many aspects of our daily lives. Silvery coatings on mirrors, antirust coatings on razor blades, antireflective coatings on camera lenses, and the metal layers on compact discs and video laser disks are all products of modern thin-film technology. On a less visible scale, the integrated circuits—microchips—that show up in all our modern electronic gear are fabricated from many layers of different thin film materials.

Josephson tunnel junctions are also made from thin films. These junctions are sandwich structures consisting of a layer of superconductor, a layer of insulator in the middle, and another layer of superconductor. We measure the thickness of such layers most conveniently in angstrom units, one of which equals ten-billionths of a centimeter, the typical size of an atom. The superconductors are generally a few thousand angstroms thick, whereas the insulator must have a thickness of only about 20 angstroms in order to permit quantum mechanical tunneling to take place. As we saw earlier, only an extraordinarily thin insulator will permit tunneling currents to flow. Thus, when we speak of thin films, we are talking about materials that are thousands or even only tens of atomic layers thick.

Scientists face many challenges in making high-quality superconducting films. To make these thin films, scientists commonly use vapor deposition techniques which often consist of simply heating up the appropriate material in a vacuum chamber until it boils away. The vapor cloud formed then strikes nearby surfaces, rapidly cools off, and solidifies as a thin film coating. This sort of process must be done in a vacuum chamber because the presence of atmosphere would result in the incorporation of oxygen, nitrogen, and other elements in the film. Generally, we want to produce the purest film of the desired substance. In order to make the purest metal films, we need the highest possible vacuum in the deposition chamber.

The real trick in making a Josephson junction is not producing the superconducting films, it is making the 20-angstrom insulating barrier. The first way it was ever done, and still the most successful technique, consists of letting nature do it for us. Most metals, when exposed to air or oxygen, form an oxide on the surface. Ordinary rust on iron provides a familiar example. In general, the thickness of such an oxide depends on the conditions under which it was formed—particularly the temperature, humidity, and oxygen content of the atmosphere. Conveniently, for Josephson junction makers, the oxide that forms naturally on metals like aluminum, tin, and lead has just the right thickness to be a tunneling barrier.

The first Josephson junctions were made in 1963 at Bell Laboratories by John Rowell and Phillip Anderson. The way they did it was to produce a thin strip of tin on a glass slide, heat it up on a hot plate in the presence of oxygen to oxidize the tin, and then deposit a strip of lead across the tin strip to complete the junction. The metal strip patterns were created by depositing the metal vapor through a mask, exactly like spray-painting letters through a stencil. The researchers then soldered electrical leads to the metal films, cooled the junction with liquid helium, and observed the Josephson effect.

In essence, all Josephson junctions are made by analogous techniques to those used in these first devices. However, there were two important drawbacks to using such early tunnel junctions for a practical electronics technology: their performance was unpredictable and they were not reliable for repeated use. The solutions to these problems came from advances in material science and vacuum technology.

Early Josephson junctions were unpredictable in several ways. Foremost was the fact that most junctions simply did not work. The insulating oxide barrier in a Josephson junction does not only need to be thin; it needs to be perfect. If it has any pinholes or thin spots, the superconductors will contact each other and there will no longer be a tunnel junction. Physically, such a device behaves as one continuous superconductor; supercurrents can flow from one side to the other. In electrical terms, the device behaves like a short circuit; current flows without any voltage. Another path to failure comes from the incredible sensitivity of the tunneling process to the thickness of the barrier.

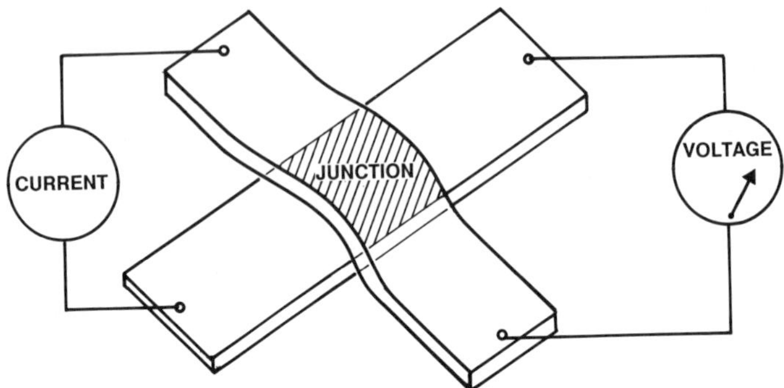

A Josephson junction formed from two crossed metal strips.

Even just a few extra angstroms of oxide can produce a junction that won't pass current at all. In truth, veterans of the early days of Josephson junctions will attest that it was amazing when junctions worked at all.

But even when scientists were able to make a Josephson junction of acceptable quality by the early techniques, it represented only a small first step toward a superconductive electronics technology. Building functioning Josephson junction circuits means producing many junctions at a time with consistent and well-determined properties. For example, earlier we learned that a Josephson junction can pass a zero-voltage supercurrent. Circuits that contain Josephson junctions are designed for a certain-sized supercurrent; any random value will not do the job. The amount of Josephson current that flows in a junction depends on the size of the junction and, very critically, on the thickness of the insulating barrier. Producing such barriers by heating the films on a hot plate, baking them in a humid oven, or other such methods, simply could not provide the control, precision, and reproducibility that a circuit technology required.

The second major difficulty with early Josephson junctions stemmed from the use of tin and lead as the superconductors. These materials melt at temperatures below 350°C and are accordingly very easy to evaporate into films. On the other hand, such films are not very durable. When researchers subject them to the stresses of being cooled down to liquid helium temperatures and warmed back up to room temperature, they tend to crinkle up, form filaments, or otherwise fall apart. Josephson junctions made from such superconductors typically worked only once, clearly an unacceptable performance level.

The solutions to both limitations of early Josephson junctions came from the application of new materials and new techniques. The interim solutions were developed by IBM in the 1970s, when they were conducting a major program to develop an entire computer using Josephson junction technology. Computer circuits require large numbers of Josephson junctions and the IBM team developed the junction fabrication process that could make them.

The process made use of lead alloys. IBM scientists found that adding small amounts of indium and gold to lead made for a much better material for making Josephson junctions. They could evaporate smoother films with accompanying smoother oxide barriers using the lead alloy technology. The IBM researchers also tackled the problems of reproducibility and control. They learned to make high-quality oxide barriers right in their vacuum system by a technique called *radio frequency plasma oxidation*. In this method, alternating electric fields actually drive oxygen ions into the films in order to form oxides at the surface. In later years, IBM supplemented the process by using niobium for the bottom electrode of the tunnel junctions instead of the lead alloy. This process therefore was based on hybrid lead-alloy-niobium Josephson junctions.

Following in the footsteps of the IBM work, modern junction makers have developed a new technology that has achieved great success in the United States, in Japan, and elsewhere. Today, the best-established Josephson junction fabrication process uses niobium films for both superconducting electrodes and aluminum oxide for the tunneling barrier. Although niobium has been around for a while, only recently has it become a routine procedure to make high-quality superconducting niobium films.

Making niobium films is challenging for two reasons: the melting point of niobium exceeds 2400°C and it has a tremendous appetite for oxygen. The high temperatures needed to evaporate niobium necessitate the use of sophisticated techniques for making films. Metals like tin and lead can be evaporated by placing slugs of metal into an electrically heated cup in the vacuum system. But these resistive boats, as they are called, cannot produce the temperatures needed to boil off niobium. Evaporating niobium requires the use of a device called an *electron-beam gun* in which very high energy electrons are fired at a chunk of niobium metal in order to heat it sufficiently to vaporize the metal. In the 1970s, niobium films were almost always electron-beam evaporated.

More recently, thin-film technologists have made niobium films using an entirely different technique called sputter deposition, which has become the preferred technique for razor-blade coatings, compact disc metallization, and a host of other commercial thin-film applications. Sputtering is not a heating process at all; it is more like atomic sandblasting. It works by shooting ions of a nonreactive gas—like argon—at a target made of the material we want to sputter—like niobium. The argon ions hit the niobium target and knock off atoms, one at a time. These atoms then scatter away and bump into the surface we are trying to coat. The pieces of glass, silicon, or other material upon which thin films are deposited are called *substrates*. When enough atoms have hit the substrate and stuck to it, we have produced a thin film. The big advantage of this technique is that the target material does not have to melt or boil; sputtering works on virtually any material.

The best present-day Josephson junctions are made from sputtered films of niobium. Sputtering neatly solves the problem of how to produce the films, but there still remains the problem of niobium's affinity for oxygen. Niobium loves to combine with oxygen and form various oxides, a process that is extremely detrimental to superconductivity; adding 1 percent oxygen to niobium reduces its transition temperature by one degree. But where does the oxygen come from in the first place? The answer is that the amount of oxygen present in a 1960s-style vacuum system constitutes more than enough to ruin the quality of a niobium film. At that time, a good vacuum system was capable of being evacuated to a gas pressure one hundred million times smaller than ordinary atmosphere. This seems like very little remaining gas but such a pressure actually corresponds to 10 atomic layers of gas hitting the system's surfaces every second. Since most thin-film techniques are hard pressed to deposit metal films very much faster than this, scientists found it difficult to make good niobium films in such vacuum systems.

Modern vacuum science has solved the problem with ultrahigh vacuum systems. In recent years, it has become practical to build thin-film deposition systems capable of being evacuated to pressures ranging from a thousand to a million times lower than those attained by the earlier systems. The development of highly efficient and clean vacuum pumps and better sealing techniques has made such perfor-

Vacuum apparatus used to fabricate niobium Josephson junctions. (Courtesy TRW)

mance commonplace. Simple-looking bell jars attached to noisy mechanical pumps have been replaced by futuristic conglomerations of stainless steel that produce today's Josephson junctions.

While the material scientists of the 1970s were learning how to make high-quality superconducting thin films, other researchers were perfecting advanced techniques for producing miniature patterns from the films. Electronic microcircuitry consists of multiple layers of microscopic thin film patterns that define various circuit elements and interconnections. The integrated circuit industry has developed advanced techniques for producing thin-film patterns based on photolithography, and such techniques are now also applied to the fabrication of Josephson junctions. Photolithography works like contact printing in a photo darkroom. To produce a pattern in a thin film, researchers coat metal films with a light-sensitive chemical called photoresist and place a glass plate printed with a pattern over the film. They then shine ultraviolet light on the combination, exposing the photoresist like ordinary photographic film. After chemical developing, the pattern has been printed in photoresist on top of the metal film. Researchers then use acids or other means to dissolve away the film, leaving behind the portions covered by the protective photoresist coating. When they finally remove the photoresist, what remains is a perfect copy of the original pattern made of thin film. Features as small as a few millionths of an inch can be printed onto thin films by these techniques. The intricate miniature features of modern integrated circuits are made in just this way.

Ever since the IBM computer project, superconducting circuitry has been made by photolithographic techniques. Furthermore, photolithography has enabled the implementation of the most significant development in junction making: the *trilayer*. Making junctions with trilayers means producing all three layers of the Josephson junction sandwich in the same operation, without any intermediate patterning of the films. This has the advantage of making all aspects of the tunnel junction process controllable and repeatable; the films never see the light of day (or the atmosphere) until the delicate three-layer structure has been completed. Trilayers are usually deposited onto 2- or 3-inch disk-shaped wafers of silicon, the same platforms upon which semiconductor integrated circuits are constructed. Researchers then use specialized photolithographic techniques to carve up parts of the trilayer into individual Josephson junctions and other superconducting circuitry. The development of the niobium trilayer with the aluminum oxide insulating layer has brought Josephson junction technology to the point where complex superconducting circuits can be reliably fabricated and even incorporated into commercial products.

Very recently, the trilayer technique has been applied to another superconductive electronics process. Scientists are now developing Josephson junction circuitry using niobium nitride, the 16 K superconductor. Making high-quality films has proven to be a tougher task using niobium nitride. Moreover, Josephson junctions must be made with entirely different insulating layers. Nevertheless, today's sputtering machines and photolithographic techniques are up to the task and niobium nitride circuits are catching up in quality with their niobium counterparts. Potential users like spacecraft designers are keenly interested in niobium nitride, because its higher transition temperature promises superconducting circuitry that can operate in advanced refrigerators that run at 10 K, eliminating the need for liquid helium.

Each new superconducting material presents a new set of problems for tunnel junction fabricators. It seems that higher transition temperatures are invariably accompanied by more complex and challenging superconducting materials. Thus only minor successes have been achieved with the A15 superconductors like niobium-tin and niobium-germanium. As we have now entered the age of true high-temperature superconductivity, the advanced materials of the 1970s may already be outdated and might never be developed for superconductive electronics.

The practical problems of making superconducting thin films and building superconducting circuits pose complex and broad-ranging obstacles. We have only touched on some of them here. Later on, we will see what can be accomplished with such circuits, regardless of the materials being employed. At this point, we shift gears entirely and focus our attention on the problems of superconducting wire, the basis of all large-scale superconducting applications.

Superconducting Wire

The most obvious way to exploit the zero-resistance property of superconductors is to build large electromagnets from superconducting wire. Permanent magnets have limited strength; really intense magnetic fields are generated by flowing lots of current through large coils of wire. Wire that doesn't heat up and requires no energy to pass current seems perfect for the job. However, years of research have demonstrated that the superconductors available to Kamerlingh Onnes and other early workers showed absolutely no promise for use in large magnets. The amount of magnetic field necessary to destroy superconductivity in lead or tin can be produced by a toy horseshoe magnet. Clearly these superconductors could not be used to build powerful magnets.

Superconducting alloys showed far greater promise. In the 1930s, researchers at Leiden found that superconductivity persisted in lead-bismuth alloys in fields one hundred times greater than the elemental superconductors could withstand. Unfortunately, some other experiments at Leiden gave the mistaken impression that such alloys could not carry large amounts of current in the presence of high fields. Scientists made these measurements at low fields and incorrectly extrapolated their results to higher fields. An entirely different kind of superconducting behavior characteristic of alloys and compounds was responsible for these misleading results.

We have already examined the thermodynamic description of the critical magnetic field in superconductors. As we saw, superconductors have to supply energy in order to ward off magnetic fields. When the energy of the superconducting state exceeds the energy of the normal state under these conditions, the superconductor reverts to the normal state. In 1957, the Soviet physicist Alexei Abrikosov showed that for certain superconductors, a different mechanism exists for dealing with this competition between superconductivity and magnetism.

Abrikosov predicted the existence of what he called the *mixed state* in superconductors. This state reduces the amount of energy that a superconductor must expend in expelling magnetic fields by letting the field partially penetrate the material. The superconductor no longer needs to push out all the magnetic field. In the ordinary Meissner effect, magnetic fields are kept at bay by diamagnetic screening currents that flow on the surface of the superconductor. In the mixed state, magnetic fields penetrate the superconductor through a number of cylindrical cores of normal material distributed across the superconductor. Tiny circulating currents surround each core, maintaining a magnetic field. These cores, known as *vortices*, each contain exactly 1 quantum of magnetic flux. As Fritz London predicted, any closed loop of superconductor can contain only certain quantized amounts of magnetic flux; a vortex in the mixed state forms just such a closed loop.

Superconductors that behave in this manner, called *Type II superconductors*, act like ordinary superconductors in the presence of small magnetic fields, but in a sufficiently large field, the remarkable mixed state takes over. In this state, the current encircling each vortex generates a magnetic field that interacts with the field of other vortices. As a result, the vortices repel one another like tiny bar magnets and they therefore arrange themselves into an orderly array called the *fluxon lattice*. To reach the lowest energy state, this lattice takes on hexagonal symmetry, much like the cells in a beehive. An applied magnetic field thus penetrates the superconductor in a regular pattern. Since each vortex contains exactly 1 flux quantum, increasing the magnetic field merely increases the number of vortices. The vortices themselves are quite tiny—their size is set by the superconducting coherence length, typically several hundred angstroms—so that huge numbers of them can fit into a piece of superconductor. The superconductor finally reenters the normal state when no more vortices can be packed into the material.

The Ginzburg-Landau theory of superconductivity provided a way to predict whether a superconductor would exhibit this Type II behavior. The key test consists of comparing the magnetic penetration depth with the superconducting coherence length. Type II superconductors always have penetration depths that are larger than their coherence lengths. How does this lead to Type II behavior? The coherence length determines the characteristic separation between the electrons in a Cooper pair. Usually this distance is so large that many pairs coexist in the same region of a superconductor. Small coherence lengths mean that Cooper pairs don't overlap much; magnetic field can therefore thread its way between Cooper pairs in discrete channels. Most elemental superconductors don't have such compact Cooper pairs, whereas niobium and most alloys and compounds do.

The penetration of magnetic fields in the form of vortices turns out to have important consequences for practical uses of superconductivity. The presence of vortices in a superconductor allows the superconductor to cope with magnetic fields in a way that requires the expense of much less energy. As a result, Type II superconductors can remain superconducting in far greater magnetic fields than their ordinary—or Type I—superconductor counterparts. This property makes these materials the prime candidates for superconducting wire applications. However, we still must ask whether Type II superconductors can carry enough current to be useful.

We have seen that in Type I superconductors, critical currents—the currents that destroy superconductivity—are related to critical fields—the magnetic fields that destroy superconductivity—according to a rule proposed by Francis Silsbee before World War I. The resistanceless state persists as long as the magnetic field generated by the flow of current does not exceed the critical field. Type II superconductors obey this kind of rule only until there is enough magnetic field to drive the material into the mixed state. At that point, an entirely different kind of behavior dominates the current-carrying ability of the superconductor.

Remarkably, the amount of current that a Type II superconductor can carry depends on how imperfect it is. The onset of resistance does not come about from the destruction of superconductivity but rather has its roots in fundamental electromagnetic principles, from a phenomenon called *flux flow resistance*.

Because of the Meissner effect, current in a Type I superconducting wire flows only at the surface of the wire. Any current flowing inside the wire would generate a magnetic field inside the superconductor, which, as we know, the superconductor will always avoid. In a Type II wire in the mixed state, however, current flows throughout the metal because a magnetic field can enter the wire's interior. According to the laws of electromagnetism, moving electrons and magnetic fields exert forces upon one another. Current-carrying wires and magnets either attract or repel one another. This force between the moving electrons in a wire and a permanent magnet produces the rotary motion in an ordinary electric motor. In a Type II superconductor, this so-called Lorentz force causes the magnetic vortices to move at right angles to the flow of current in the wire. The motion of the normal cores through the material dissipates energy in the form of heat; the cooperative motion of Cooper pairs plays no role here. Since passing current consumes energy, the wire suddenly develops resistance. The superconductor can no longer sustain current without using energy.

This mechanism of energy dissipation known as flux flow resistance results in very small critical currents for Type II superconductors. As soon as the vortices start moving, the zero-resistance advantage of superconductivity is lost. Fortunately for users of superconducting wire, scientists have found a cure for this problem.

Flux-flow resistance can be eliminated by pinning the vortices in place. How do we do that? Engineers purposely introduce imperfections into the wire by adding extra elements to the material or by mechanically stressing the wire during the process of drawing it into its final form. Deviations from the orderly arrangement of atoms in the metal act as pinning centers for vortices, just as imperfections in a crystal act as obstacles to electrons that increase ordinary electrical resistance. To make things easier, not all the vortices have to be pinned. Once a reasonable number of vortices are trapped in place, others get constrained by the repulsive forces between vortices and the whole vortex lattice becomes immobilized. At that point, substantial currents are needed to break loose the vortices, and the critical current of the superconductor becomes quite sizable. Thus, making the most flawless material does not produce the best possible wire.

Practical superconducting wire must be manufactured to be imperfect in a controlled fashion.

Superconductivity researchers learned these lessons in the 1960s. A group at the Bell Telephone Laboratories was able to construct a powerful superconducting magnet from the molybdenum-rhenium alloy that Bernd Matthias discovered a few years earlier. Eugene Kunzler at Bell demonstrated high currents in the presence of strong magnetic fields in yet another Matthias material: niobium-tin. He produced wire from this compound by compacting superconducting powder into a metal tube, and then drawing it into wire. In 1961, Bell Labs thought it had found the ideal wire material, a niobium-zirconium alloy, but by the next year, the real workhorse of the superconducting wire business had been found.

By 1962, a team headed by Ted Berlincourt at a company called Atomics International found that an alloy of niobium and titanium had superior properties for superconducting magnet applications. The material has a simple crystal structure and has very good mechanical properties on account of the mechanical similarities between niobium and titanium. Niobium-titanium can easily be shaped and drawn into wire—wire makers use the term *ductile*—and it superconducts above 10 K. It also meets the needs of superconducting magnet designers as it remains superconducting in magnetic fields a thousand times stronger than elemental superconductors can withstand. This kind of wire has now become an established commercial product and can be found at work in a diversity of applications in science, medicine, and industry.

In recent years, the hunger for even greater magnetic fields has spurred the development of other superconducting wire materials. Another A15 superconductor, vanadium-gallium, has become a practical material for the manufacture of superconducting wire and can stay superconducting in truly prodigious amounts of magnetic field.

Each new superconductor requires new techniques for wire production. Wire made from A15 superconductors can't be made by simple mechanical techniques because producing the superconductors requires special processing. Techniques must employ rather exotic means to incorporate the complicated metastable crystal structures into wires. Years of hard work precede the development of such specialized techniques, which often end up with colorful names like the "modified jellyroll process."

Even when the superconductor itself can be simply made, as is the case for the niobium-titanium alloy that dominates the superconducting wire business, engineers face tough design problems in producing practical superconducting wire. We will come back to some of these problems shortly. Whatever the details of the wire manufacturing process might be, the physics of Type II superconductivity has led to a practical technology based on superconducting wire. Soon we will see how such wire gets put to work.

Before leaving the topic of practical superconducting materials, we should mention that not every application of superconductivity requires thin films or wire. A number of important uses for superconductors utilize solid pieces of superconductor. For instance, solid spheres of niobium that float in a magnetic field are being used as low-friction bearings. Such bearings will find their way into satellites in the form of ultraprecise gyroscopes that provide guidance for advanced measurement devices. The diamagnetism of superconductors gets exploited in other ways as well, particularly for the purpose of providing magnetic shielding.

Superconducting magnetic shielding—which uses the superconductor's ability to exclude magnetic fields—can be exploited in a number of applications. For example, engineers require the complete exclusion of magnetic field that typifies Type I superconductors when ultrasensitive magnetic instruments are being used. SQUID (Superconducting Quantum Interference Device) instruments used to study tiny magnetic signals in the brain and elsewhere would be useless without superconducting magnetic shields. The noisy signals emanating from the changing magnetic environment of the outside world would drown out the minuscule signals that SQUIDs can detect. Another application of superconducting shields comes from the need for security in computer installations. Functioning computers generate signals that can be detected by sensitive enough equipment. Computers that handle confidential information could be protected from such snooping by superconducting shields.

At the opposite extreme, giant superconducting magnets generate stray magnetic fields that can interfere with the functioning of sensitive electronic instrumentation. Superconducting shields made from Type II superconductors can screen out the bulk of these unwanted fields. For these applications, A15 superconductors like niobium-tin are particularly desirable, since magnetic screening—like other superconducting properties—becomes strongest at small fractions of the critical temperature.

So scientists have developed uses for many forms of superconducting materials. Each form has prompted its own path of research and development accompanied by novel machines and methods appropriate for the manufacture of useful superconducting materials. Discoveries of new and better superconductors have always initiated new development efforts aimed at bringing the new materials out of the laboratory and into the working world. We return to this very subject later when we explore the challenges of the new, high-temperature superconductors.

At this point, we have come to understand some of what superconductivity is all about and something of what goes into making superconductors useful. We now turn to the topic of how superconductors can be used. The applications of superconductivity span across many industries and interests and require the expertise of workers in diverse disciplines. Most of the applications make use of the products of thin-film experts and wire developers, but how these materials can be used is only limited by human imagination. We start our survey of superconducting applications with the uses for superconducting wire.

CHAPTER 7

Miles of Wire

What is the most expensive piece of equipment you have used today? You probably used it in every room of your house. You used it at work, while you were shopping, and at play. Do you need another hint? It costs several billion dollars, but luckily you get to share it with quite a few of your neighbors. What is it? The electrical power system, a complex array of equipment that, for the most part, we can take totally for granted.

The enormous electrical power system we casually activate with the flick of a light switch constitutes a $100-billion industry in the United States. The electric utility companies have the job of acquiring sources of energy, converting that energy to electricity, manipulating the electricity into useful forms, and delivering it conveniently and inexpensively to our electrical outlets. To face this responsibility, the utility companies are constantly on the lookout for new technologies to improve the system's performance, especially in terms of reliability and cost. One technology getting deservedly serious scrutiny is superconductivity.

There are many and varied potential roles for superconductivity in the modern electrical system. In some jobs superconductivity will slowly replace normal conductors. The basic machinery will be the same, only it will perform better with zero-resistance wire. In other cases superconductivity will enable engineers to use entirely new approaches to supplying electrical energy. Machines and devices that would be inconceivable with conventional wiring may well change the very nature of the electrical utility system.

Before we can go out and build superconducting generators, energy storage systems, and power transmission lines, we need to produce large quantities of superconducting cable. Nowadays, one can buy superconducting wire from specialized manufacturers, but this has not been the case for long. Making superconducting wire that meets the needs of the energy industry is no simple task. There were fundamental physical and engineering problems that had to be solved for every superconductor currently in use, and will have to be solved for the superconductors of the future. Before we examine the role of superconductivity in the power industry of the future, we examine some of the important issues in the making of superconducting wire.

We know in principle that superconductors can carry tremendous amounts of current through relatively tiny wires. Superconducting wire capable of handling one hundred amperes, enough current to supply a family home, is not much thicker than a human hair. By contrast, copper wiring for the same power level must be nearly an inch across to avoid overheating. Engineers are not primarily interested in exploiting this capability for individual homes, but rather for large-scale electrical power plants and power lines.

Why should they care about high currents in tiny wires? Why not just use very big wires made of copper or aluminum to deliver electrical power and forget about superconductivity? There are a number of reasons and most of them center around economics. Copper and aluminum are fairly cheap, but when utilities install hundreds of miles of cable at a time, the material costs start to add up. Thinner wire means less mining, refining, and manufacturing costs. There are other advantages to high-current-density wires as well. For many electrical power applications we have in mind, superconducting wire offers the combination of compactness and resistanceless current flow that permit us to perform technological tricks that would be impossible with ordinary conductors.

Providing enough electrical power to supply a single home is easy. The big technical problems come in supplying power to millions of households at the same time. Zero resistance and high-current capacity give superconductors clear-cut advantages in handling our enormous electrical power needs. That's a good starting point. However, the real success or failure of superconducting power cable will come from engineers' ability to meet some of the other requirements for the job.

When utility engineers write out their list of performance criteria for electrical cable, low resistance is only the first of many items. Manufacturers of superconducting cable must meet a long series of electrical and mechanical requirements before their product can be deemed acceptable. Ideally, cables should be resistanceless, carry very high currents, and operate at practical temperatures. Since large currents generate intense magnetic fields, superconducting cable must withstand large magnetic fields and still maintain full superconducting properties. Apart from its electrical performance, superconducting cable must be rugged, flexible, and convenient to refrigerate. In addition, the cable must be practical to manufacture in large quantities at affordable prices. After years of intense effort, the builders of superconducting cable routinely meet these requirements in producing today's high-current-density, high-magnetic-field wire.

Superconducting wire is now a well-developed commercial product. Such wire bears little resemblance to simple strands of superconducting material. Considerable engineering and manufacturing skill goes into the production of practical superconducting wire. We can get a feel for the strides wire manufacturers have made over the years by looking at one problem that pretty much has been licked: the *quench*.

The quench is a funny-sounding malady that used to be annoyingly common in large superconducting magnets and power cables. Like a medical ailment, the quench has distinct symptoms, a root cause, and, luckily, a cure. The symptoms of a quench in a large superconducting magnet border on the spectacular. One minute the magnet merrily does its job, conducting supercurrent and generating large magnetic fields while its helium bath bubbles quietly. The next minute the wire has all turned normal, the formerly superconducting currents convert to normal current, and the resultant heating rapidly boils all of the helium away in an explosive plume.

Engineers quickly traced the cause of quenching to the cable itself. The earliest designs for superconducting cable had no tolerance for temporary losses of superconductivity with the wire. Such an event could happen as a result of lightning-induced current surges, transient fields generated as the coil is energized, bad spots in the wire, or frictional heating as wires shift around. Whatever the mechanism, if the superconducting path is temporarily interrupted, even in just a small segment of the wire, then the cable is in for trouble. Not only will that segment of wire heat up, but the heating will in turn destroy superconductivity in adjacent regions as well. Soon a thermal avalanche starts with no way to stop it.

Superconducting cable acts like a super highway for electrons, and quenches are like a jackknifed truck that ties up the entire roadway. Just as one crippled truck can render a major traffic artery useless in a matter of moments, the destruction of superconductivity in one spot in a superconducting cable can bring about its demise. How can such catastrophes be averted?

We have yet to learn how to eliminate all accidents on our highways and, similarly, superconducting cable manufacturers cannot avoid all the possible triggers for quenches. Instead, they follow the example set by the highway system. When an accident ties up a roadway, drivers are forced to take an alternate route. To follow this highway analogy, cable manufacturers have devised rather clever defenses against quenches. The trick is to coat the superconducting wire with fairly thick copper. If the superconductor warms into the normal state, the copper can carry the diverted current with only minimal heating. Why do we need the copper? Because the superconductor is a poor ordinary conductor in its normal state. Eventually, when the superconductor recovers, it again takes over the current burden and the wire returns to its zero-resistance state. In addition, copper conducts heat away from the superconductor so readily that prolonged temperature rises become less likely. Additional tricks such as combining many microscopic twisted superconducting wires in a copper-clad bundle rather than relying upon a single superconducting pathway lead to a network of safety channels for current flow.

In terms of our traffic analogy, the copper acts as a wide bypass road. Any time the need arises, traffic can detour onto the alternate electron pathway. Even though the quality of the road pales in comparison with the superconducting highway, the extra avenue prevents the big pileup at the weak spot. The technique results in a "cryostabilized" wire that is largely immune to the annoying quench failures of early superconducting wires.

Engineers produce cryostabilized cable by surrounding superconducting filaments with normal metal in order to form a composite structure. In ordinary current flow, the normal metal has little function, since current always takes the path of least resistance through the superconducting filaments.

Cross-sectional view of a composite wire consisting of 93 niobium-titanium wires embedded in copper. (Courtesy Supercon, Inc.)

The normal metal in superconducting cable does play an important part in losses that occur when alternating currents flow through superconductors. Direct currents generate constant magnetic fields around superconducting cable. Such magnetic field have no ill effects and the cable carries the current with no resistance. Alternating current, on the other hand, even at low (60 Hz) power-line frequencies, brings significant loss mechanisms into play. The culprit is the changing magnetic fields generated in the superconducting cable. Recall that for Type II materials—like the niobium-titanium alloy used in superconducting cable—magnetic fields can penetrate throughout the superconductor. Alternating currents therefore produce changing magnetic fields within the cable. These changing fields in turn produce reactive currents, descriptively known as *eddy currents*, within the normal metal part of the cable. These flowing eddy currents in the normal metal dissipate energy in the form of Joule heating, thereby producing a noticeable resistance in the superconducting cable.

Engineers minimize eddy current losses by providing large numbers of superconducting filaments along a sharply twisting pitch through the cable. In addition, the engineers intentionally produce high-resistance regions in the normal metal in order to promote the rapid decay of the eddy currents and minimum Joule heating losses. Superconducting cable manufacturers have worked very hard to combat problems such as the quench and eddy current losses, and the payoff may well be reaped in every segment of the electrical utility market. High-quality, stable superconducting cable forms the cornerstone of new technologies for the electric power industry. While superconductivity may well affect all phases of the utility industry, perhaps the most straightforward applications will come in electric power generation.

Producing Power

The hunt for more efficient ways of generating electricity is more than an exercise in saving a few pennies on our fuel bills. Power companies are running a race against time and they know it. Supplies of oil and coal, our fossil-fuel heritage from the time of the dinosaurs, won't last forever. We have only a limited time to develop renewable energy sources as well as to improve the efficiency of the existing energy technology in order to keep up with our electrical power needs.

Generating electricity hasn't changed much since the days of Thomas Edison. One way or another, power engineers convert mechanical motion into electricity. The machines they use, called *dynamos* or simply *generators*, rely on the motion of wires cutting through the fields of enormous magnets to create electrical power.

Although the mechanism for generating electricity is simple, the logistics are not. One electrical power plant may produce energy for a million people. The generators to serve these needs are gargantuan. Efficiency and economics dictate the use of a few big generators instead of many smaller, neighborhood plants. Quite often the ultimate generator size is limited by the ability of generator magnets to fit through railroad tunnels on the way to the construction site.

The superconducting solution to the generator-size problem replaces massive and voluminous iron magnets with lighter, more powerful superconducting magnets. Thanks to the intense magnetic fields produced by superconducting magnets, generators of comparable or greater power capacity can be much more compact. Engineers also like the light weight of superconducting systems, which enhances both their mechanical and electrical stability. Laboratory demonstrations of prototype systems have been successful, and the recent advances in superconducting materials have stimulated interest in new large-scale tests of superconducting generating systems.

Better generators are not the whole story for superconductivity and the production of electricity. While superconducting generators may replace their conventional technology counterparts in the moderately near future, far-sighted scientists are already working on even better ways of producing electricity. There is growing excitement in the scientific community over ways to generate electricity that would be relegated to the realm of interesting but impractical science fiction without the help of superconductivity.

One new method for producing electrical power goes by the rather unwieldy name of "Magneto Hydro Dynamic (MHD) generation." These generators have been the subject of numerous investigations by power engineers over the past two decades. The reason for all the activity is simple: MHD uses less fuel to create the same amount of electricity as conventional generators. Furthermore MHD generates power with almost no moving parts. How does MHD work?

An MHD generator starts with burning fuel to create a very hot gas. At sufficiently high temperatures some of the electrons are stripped away from their atoms because of highly energetic atomic collisions. The result is a "plasma," a combination of neutral atoms, charged atoms, and electrons coexisting in a flamelike gaseous state. In the MHD generator, the plasma travels down a special flow tube surrounded by powerful superconducting magnets. Only superconducting magnets can economically fill the flow tube with magnetic fields strong enough for efficient MHD power generation. The magnetic field causes positively charged atoms to deflect into one wall of the tube and pushes electrons into the opposite wall. Charge collectors along these walls directly gather electrical current from the hot gas. In this way, energy of the flowing gas is converted directly into electricity without the complexity of turbine blades, rotating magnets, or the massive dynamos of conventional power generation.

Sometimes two heads are better than one, and MHD combined with conventional generators take advantage of this principle. The power industry is studying MHD systems as "topping" units mated with conventional power plants. The pairing of MHD with a conventional generator forms a powerful team. The power generation from the symbiotic pair begins with hot gas from a coal fire. An MHD stage efficiently pulls electricity directly from the gas, but abandons large amounts of untapped heat to the exiting gas. Instead of wasting the valuable energy contained in this exhaust gas, the two-stage system directs the hot gas to a set of boilers, just as in conventional power plants. Steam from the boilers turns turbines and powers conventional generators to tap the remaining fossil fuel energy. In essence, two generating plants get to convert the same fuel to electricity. By converting heat to electricity in two ways, the combination topping plants excel in fuel efficiency over single method systems. Engineers have high hopes that the fuel savings from such topping systems will more than pay for the investment in the MHD equipment.

Superconductivity has a part to play in even more futuristic energy systems. One approach proposes to extract energy from water. It sounds incredible but that is where a number of scientists are directing their investigations to solve our energy problems. Hydrogen and oxygen form water, and it is the hydrogen that scientists see as an answer to our future energy needs. A primal process in the universe consists of the fusing together of the nuclei of two hydrogen atoms to form a helium atom. When this "fusion" process occurs, tremendous energy is released. Scientists and many decision makers are betting on fusion power to solve the looming energy crisis of the future. Fusion means almost limitless power. It has fueled the sun for 5 billion years and will continue to do so for a few billion more to come. The same energy source is unleashed in the awesome violence of the hydrogen bomb, but a bomb represents an uncontrolled release of energy. Taming the power of nuclear fusion to provide electricity is a Herculean task.

In the sun, fusion occurs under the conditions of million-degree temperatures and tremendous pressures. On earth, the conditions have most easily been duplicated with the explosion of an atomic bomb. The nuclear fusion in future power plants will require ignition conditions similar to the atom bomb or the interior of the sun, but maintained in a controlled generating plant. How does one contemplate accomplishing this?

Imagine holding a lightning bolt in a bottle. That is exactly what fusion scientists try to do, except that their lightning bolts are thousands of times more energetic than anything found in a summer thunderstorm. The container the fusion physicists use is not an ordinary bottle, but a magnetic one. Not even million-degree plasmas can escape from the grip of a sufficiently intense magnetic field. Extremely powerful magnets can act to channel lightning-like plasmas back onto themselves and only superconducting magnets are powerful enough to do the job. No conventional magnet even comes close to providing the gigantic magnetic fields and house-sized volumes needed to contain the forces of the sun.

Scientists admit they are decades away from fusion energy plants, but the battle goes on. In the meantime, we will concentrate on using our fossil fuels as efficiently as possible by developing superconducting generators and MHD with superconducting magnets. Magnetically confined fusion will provide us with power after the exhaustion of our fossil fuel reserves. But whether electrical power comes to us from fossil fuels, nuclear fusion, or alternative energy sources, generation of electricity is only the first step in the electrical utility system.

Storing Energy

Power engineers study energy storage as a key ingredient in electrical supply grids. Why is electrical storage the second major segment of our electrical supply system? Why bother storing energy? Why not just generate electricity and supply it immediately to the consumer? The answer has to do with supply and demand because when it comes to electricity, the two can't always keep in step.

The supply side of the electricity equation is simple. Power plants operate most efficiently at constant full power. Fuel efficiency is highest and labor costs are lowest for power plants putting out full power. Financial considerations push for constant full power output as well. It costs the same large amount of money to build an operating power plant as it does to build an idle one. An idle power plant, or one operating at a fraction of capacity, wastes capital resources. Utilities must make the same loan payments on utility plant equipment whether or not the generators are running.

As consumers, our needs differ from those of the power companies. We demand power to suit our needs. On hot days we air condition our buildings. On cool nights we rely on electric heating.

Our requirements constantly fluctuate according to climate, industrial activity, and our personal whims. Despite the changing nature of our needs, we expect the utility companies to provide us with electrical power no matter how many other people require power at the same time.

To deal with such a fundamental conflict, power companies are forced to follow a worst-case building program. The industry must always establish enough generating capacity to satisfy the largest conceivable instantaneous demand of all its industrial and residential customers. Power companies build large and expensive plants knowing that the generators will sit idle much of the time. These costs are quickly passed on to the consumer in the form of increased utility bills.

As power companies exploit new, alternative energy sources, the supply–demand mismatch worsens. Solar energy may eventually supply a large fraction of our electrical requirements, but solar collectors will only work during the day. Harnessing the energy of ocean tides has also excited interest, but tidal power only comes in a series of surges. Wind power, tapped by giant turbine blades, can be equally unsteady. These pollutionless energy sources will no doubt play a greater role in energy generation. However, they leave utility engineers with the problem of finding a practical way of matching the availability of this renewable power to consumers' power needs at any time.

Two strategies confront the power companies. They can build quick-response power plants, generally using expensive coal or oil, to meet peak demands. This path costs utilities, and therefore consumers, large capital outlays for underutilized equipment. Alternatively, the power companies can fashion some scheme to store excess energy during low-power-requirement periods for use during peak demand. Energy storage can be the most economical answer to a cumbersome problem. This is why engineers are trying to develop economical energy storage technologies.

Some energy storage systems are already in use. Hydroelectric plants store water in reservoirs, releasing water to the generator turbines as needed. Pumped water storage systems take the concept of energy storage one step further by actually carrying water uphill into reservoirs when surplus power is available. Several problems plague these energy storage techniques. Both hydroelectric variations of energy storage require huge amounts of land, good water-retaining geography, and plentiful supplies of water. Power engineers still search for easier ways to store energy.

Superconductivity may be the solution to the electrical-energy storage problem. Engineers are investigating burying football-field-sized superconducting coils in the ground to store energy in an electrical form. In the industry, these systems are called SMES, for Superconducting Magnetic Energy Storage. As the name implies, practical electrical power storage is made possible with the lossless current flow available in superconducting wire.

Operation of superconducting magnetic energy storage is straightforward. When excess electrical energy is available, current is fed into a superconducting coil. Once the current starts, it will flow undiminished in a closed superconducting coil. When extra power is needed to satisfy customers' electrical demands, current can be diverted out of the coil and back into the power system. In essence, the superconducting coil acts like a savings bank for electrical current. We put in deposits when we have something to spare and make withdrawals during times of need.

At first glance, it may seem paradoxical that superconductivity can excel in energy storage. After all, many of the most important applications of superconductivity are based on the perfectly lossless flow of current through a superconductor. How can the current that takes no energy to maintain be harnessed so effectively to store energy?

The answer is that there can be a great deal of energy contained in the current in a superconducting coil even though the coil uses no energy in carrying electricity. The trick is to store the energy in a giant inductor, which is actually just a coil of wire. Inductors are the inertial element in electronic circuits. What does that mean?

Electrical circuits have analogues in mechanical systems. In particular, circuits can contain kinetic and potential energy and even the property of inertia, the resistance to change. Kinetic energy—the property of moving objects—shows up in the form of current. Potential energy—like the energy stored in a compressed spring—is found in capacitors in the form of stored charge and in inductors in the form of magnetic fields. In mechanical systems, inertia resides in mass. A massive object is difficult to move and, once it is moving, is difficult to stop. An object with a lot of inertia wants to hold on to its kinetic energy. In electrical circuits, inertia is a property of inductors, which are most simply coils of wire. The inertial property manifests itself in the fact that it is difficult to establish a current in a large inductor and, once this is done, the current is difficult to stop. Analogous to a heavy moving object, a large inductor hangs on to its energy. However, in an inductor made of ordinary wire, there is another place for the energy to go: Joule heating. Pumping huge amounts of current into a conventional coil is mostly a good way to heat the coil. If we want long-term energy storage, we need a superconducting coil.

Such coils have tremendous electrical inertia, hence a large superconducting coil takes little or no energy to maintain a current flow, but may take several hours' output from a generating plant to reach its full current capacity. Where does all the energy in a superconducting coil go? Magnetic field energy takes the place of the energy of motion in a mechanical system. Silent, invisible, penetrating, and motionless, magnetic fields can still be highly energetic. The magnetic field we are most familiar with, the earth's field, doesn't reveal its magnetic energy because the fields are relatively weak and are quite steady. The magnetic fields possible with superconducting coils are billions of times more energetic and are easily tapped for our use. We can access the magnetic energies of superconducting coils by controlling the currents circulating around the coil windings.

SMES designs aim for maximum possible energy storage. The total magnetic field energy depends on two quantities: the magnetic field strength and the volume of space encompassed. SMES for electrical power systems involves the use of extremely large magnetic fields, which are the speciality of superconductivity. Power system engineers have examined a few coil layouts, and generally prefer a large superconducting ring with many turns of wire. Such a system could be built economically with standard construction techniques, although there are complications to deal with.

A serious consideration for such designs is that powerful mechanical forces act on SMES coils. We know that magnetic fields exert force upon current-carrying wires. We depend on such magnetic forces in the operation of electric motors. In the case of SMES, the forces are truly something to contend with. The combination of enormous currents and gigantic magnetic fields present in SMES leads to expansive pressure along the coil's frame. As a result, designers need to anchor the SMES unit firmly in bedrock for mechanical support. Without such firm support, SMES coils would literally tear themselves apart.

The financial implications of SMES units could change the face of the utility industry. The coils envisioned for large-scale SMES could absorb or discharge the output of a large generating plant for an entire day. Such power-storing abilities could save power companies millions of dollars. The least efficient generating plants, comprising about 15 percent of the generator work force used only to satisfy peak demands, could be decommissioned. In their place, utilities would run their most efficient plants full-time, even during periods of low demand. Excess energy from these off-peak hours would be SMES-stored for delivery at peak demand times.

There are other applications for magnetic energy storage that go beyond the simple daily averaging of electrical power requirements. Sometimes energy storage for only a few minutes is desirable. Consider what happens if a power plant suffers a mechanical failure. It takes time to turn on replacement generators. A charged magnetic field coil could supply power at an instant's notice. A smaller, building-sized SMES could replace a generator for several minutes, enough time to turn on replacement generators, tap additional power sources, or reduce the load gracefully.

Sometimes utility engineers find themselves fighting the battle in reverse. When electrical power demands drop suddenly, or an alternative energy source comes on line, it takes time to reduce power levels at conventional generating plants without wasting energy. Magnetic coil storage can usefully absorb temporary excesses of power for later use.

Furthermore, SMES can tackle jobs where enormous amounts of energy need to be stored for only seconds. The first large scale SMES unit attached to an American utility did just that. Bonneville Power Authority utility engineers were having problems in a very large power grid connecting their giant hydroelectric plant in Oregon to users in Southern California. They had to find a way to steady power levels that were swinging up and down every few seconds. A superconducting storage coil, built by Los Alamos National Laboratories, successfully handled the job, stabilizing the utility supply system. Whenever the system voltage would rise too high, current would be bled off into the superconducting coil. When voltage started to drop, current would be withdrawn from the coil. SMES gave the utility engineers a way to regain control of their power grid.

Superconducting energy storage can also be used in systems requiring enormous amounts of energy for even shorter amounts of time. Fusion generators require just such an energy source. The million-degree temperatures required for thermonuclear ignition require tremendous rapid heating. Energy can be slowly accumulated in a magnetic storage system and then quickly expended to initiate nuclear fusion.

Electronic warfare also has requirements for the enormous short duration power levels that SMES systems can deliver. Radar jammers and other electronic countermeasures use tremendous power levels during brief transmission times. Similarly, proposed laser weapons have gigantic energy requirements for their brief shots. A significant part of the Strategic Defense Initiative (SDI, or "Star Wars") planning revolves around such "directed energy" weapons capable of destroying incoming threats using ground-based and space-based energy sources. Deployment of such systems is years away, but adequate power plays an important part in system designs. Superconducting magnetic energy storage can provide the tremendous peak power output levels required to make these ideas feasible.

It seems like the uses for efficient energy storage systems are as numerous as the applications requiring large electrical power. No matter what the application, superconducting magnetic energy storage is the key to handling energies far beyond the scope of competing technologies. Thanks to superconductivity, the very nature of how we manipulate electricity between generation and delivery will change.

Superconducting Power Lines

The third key element in electrical power systems is delivery of electricity to the customer. Since the earliest days of superconductivity, the idea of superconducting wire carrying power losslessly from power plants to consumers has excited the imagination of electrical power engineers. Over the years, there have been numerous experimental investigations of superconducting power lines.

Electrical power transmission constitutes an integral and rather visible part of the electrical utility system. Our society is very dependent on plentiful, inexpensive electrical power delivered upon demand at every electrical outlet around us. The economics of generating power dictate large, central power plants rather than a multitude of smaller local ones. In addition, the logistical requirements of power plants—plenty of cooling water, access to fuel supplies, and suitable evacuation procedures for nuclear plants—also dictate that we separate the power sources from the consumers by sizable distances. Everybody wants electrical power but nobody wants a power plant in his or her backyard.

Electrical transmission lines perform the task of transporting energy from the generator to the consumer. Proponents of superconductivity propose two basic avenues of penetration into the job of electrical power transmission: long-haul power transmission and urban underground cable. We first examine the viability of superconductivity for high-power, long-haul electrical power transmission.

Such transmission lines are found leading from favorable hydroelectric sites, like Niagara Falls or Boulder Dam, to major cities. Electrical power transmission is a mature technology, having slowly evolved under the forces of the marketplace over the past 100 years. The question is whether superconductivity has sufficient economic attractiveness to find a niche in this important field.

The interest in long-haul power transmission using superconducting cable centers around avoiding the inevitable wire-heating losses experienced in high-tension wires. If superconductivity could succeed in making electrical power transmission over more than a couple hundred miles economically attractive, then the entire way utilities are run will change. Nuclear plants could be located very far from population centers, making emergency evacuation plans simpler and increasing human safety. Oil-burning plants could be located along the coasts where the oil arrives by tanker. Not only would there be plenty of ocean water available for power-plant cooling, but the costs of shipping the oil to power plants serving inland cities would be eliminated.

Even before the new high-temperature ceramic superconductors appeared on the scene, power engineers had a pretty clear picture of what a superconducting electrical transmission system would look like. Heavy superconducting cables would be run through an underground insulated pipe. The pipe would circulate liquid helium to cool the cable to superconducting temperatures. Periodically spaced refrigeration plants would remove heat leaking into the conduit or generated by losses in the wire. A number of laboratories designed and tested cable suitable for high-power transmission. Scientists conducted tests of cable sections hundreds of meters long at a specially constructed test site at Brookhaven National Laboratories in New York.

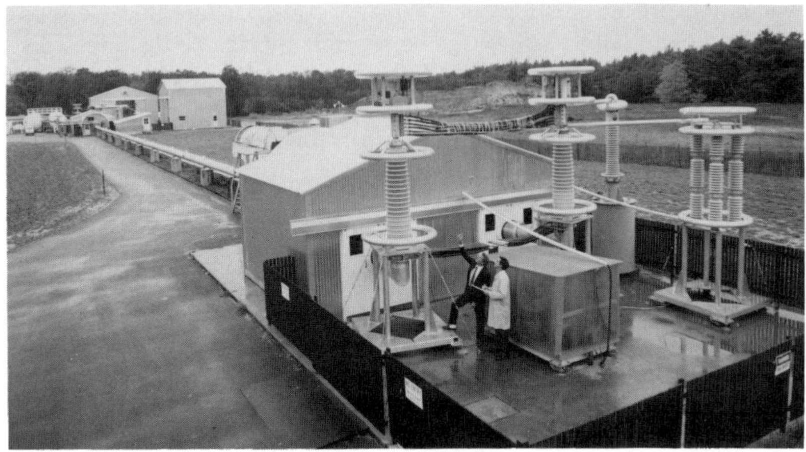

Test facility for evaluating superconducting transmission cable. (Courtesy Brookhaven National Laboratories)

The results of tests and subsequent studies showed that at present long-haul superconducting electrical transmission is technically possible but economically unattractive. Significant progress in high-temperature superconductors will be necessary before power companies seriously consider converting to superconducting lines for long-haul transmission. Although the superconducting wire technology exists today to handle the power levels of even the largest nuclear power plant, economics dictate against immediate application of superconductivity in this field. The superconducting cable itself losslessly carries even enormous currents, but systems engineers rightly point out that refrigeration costs cannot be ignored in the energy equations. When the costs of refrigeration using liquid helium are added in, present cables do not have a clear efficiency advantage over very high voltage overhead power lines. If cable can be produced using the new superconductors that require only liquid nitrogen cooling, then the tables may turn in favor of superconducting power transmission.

Decades of development have produced some alternatives to superconducting cable. By boosting transmission voltages close to a million volts, the amount of current required to transport electrical power is drastically reduced. In this way, power companies can maintain overhead line losses of only a few percent for every 100 miles of transmission line. Given the established system of overhead transmission of electricity and the low costs of erecting the transmission towers and power lines, utilities remain unwilling to risk the investments involved in laying superconducting power line.

Nevertheless, even a savings of a single percent in electrical power transmission efficiency translates to savings of a billion dollars a year in fuel and building expenses. The most sophisticated and expensive superconducting system could eventually pay for itself with such savings. Advances in the new high-temperature superconductors, and certainly the development of even higher temperature ones, would bring about a resurgence of interest in this field.

Power engineers are more optimistic about the near-term use of underground superconducting cable over shorter distances. Superconductivity may fill an important market niche for delivery of power using buried cables over distances of 20 to 30 miles. As real estate costs rise, and suburban areas expand around cities, people are unwilling to have high-voltage power lines cut across their communities. Complaints about TV reception, aesthetics, and possible environmental damage are forcing utilities to consider underground cable as an alternative to overhead power transmission lines in suburban and metropolitan areas.

For underground power transmission, superconducting cable has major advantages. Since utilities must commit to the costs of underground conduit anyway, the expenses of a thermally insulated pipe are not as dominant. Heating in ordinary cable limits its power handling and efficiency. The high current and low loss of superconducting cable may pay for itself in the long run.

Whether niobium-alloy or ceramic superconductors make the leap to commercial viability first in the coming decades, superconducting cable will appear first in short urban runs. As the technology improves, long-haul electrical transmission may convert to superconductivity as well.

On a final note, engineers have studied the use of superconducting switches for protecting utility systems from overloads and faults. The electric utility grid handles massive amounts of power produced at many power plants. Overloads, downed power lines, lightning strikes, and short circuits all have the capacity to damage power-generating equipment and transmission systems. To control such potential calamities, designers need a way to switch off the enormous currents of the utility lines whenever trouble occurs. Such switches

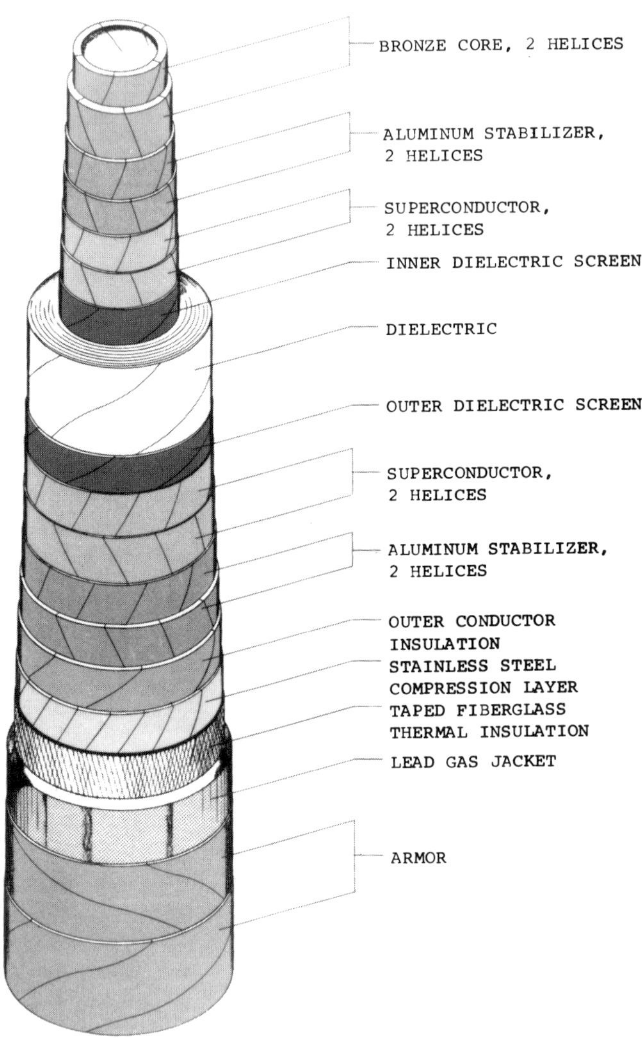

Superconducting power transmission cable. (Courtesy Brookhaven National Laboratories)

must not only be able to control huge amounts of power, but also must not needlessly waste power during normal operation.

Researchers have performed initial studies on the use of specially patterned superconducting films as power line switches. In the superconducting state, the films pass current with zero resistance; in the normal state, the films are prepared so as to have high resistance. The high-resistance state functions as the "off" position of the switch, effectively cutting off the flow of electricity. When the switch is "on," the superconducting state subjects the power grid to no electrical loss during normal operation. Apart from their electrical power applications, superconducting switches have aroused the interest of engineers in a number of other fields who need a way to switch large amounts of power in a small amount of time.

Large-scale superconductivity seems naturally suited to a wide range of tasks facing our electrical power utility companies. Generation, storage, and transmission of power will all benefit from the performance that only superconductivity can provide. The list of new technologies, magneto-hydrodynamic generators, fusion, superconducting magnetic energy storage, and the like will continue to grow as scientists and engineers explore the new possibilities presented by miles of superconducting wire.

CHAPTER 8

On the Move

Levitation and Transportation

Understanding superconductivity involves studying the interaction of electrons and matter on a microscopic scale. Furthermore, many superconducting applications involve unfamiliar concepts related to sophisticated electronics and the detection of electromagnetic fields. But more of our experience with the world centers around macroscopic motion and the action of machinery than it does on electronics, and no machines interest us more than the ones in which we travel. So at this point, we explore superconductivity's role in transportation.

When it comes to transportation, superconducting magnets will play the starring role. For most of us, our only direct experience with magnetism comes from small-scale magnetic gadgetry, like the little magnets that we stick to our refrigerator doors. But magnetism plays a far more important role in the modern world. Magnetism translates the motion of electrons in wires into the motion of large-scale objects. Magnetic forces give electric motors their drive, and to make powerful motors, you need strong magnets.

Magnets that perform the tough jobs are always electromagnets, made by flowing electrical current through coils of wire. Permanent magnets can be strong for their size, but they can't, for example, carry cars around junkyards. So electromagnet builders look for ways to build stronger magnets with better kinds of wire. It comes as no surprise that superconducting wire, with its resistanceless flow of electricity, enables new magnet applications beyond the scope of conventional electromagnets.

Suppose we were given the job of building one of the world's strongest magnets. How would we proceed? No permanent magnet—like the sort we stick to refrigerator doors—can be as strong as the most powerful electromagnet, so we would use an electromagnet design. Electromagnets are made by winding a big coil of wire. The field strength in an electromagnet—which determines the magnetic force it can apply—comes from the number of ampere turns it contains. To get more field, we can either flow more current (amperes) through the magnet, or add more turns of wire. In general, it makes sense to use as many turns of wire as possible in order to get the most magnetic field possible from the drive current. Normally, we would use copper wire to wind a magnet, at least until the advent of high-field superconductors in the 1960s.

Once we have built our electromagnet, we need a power supply to operate it. In addition to providing enough current to generate the desired magnetic field, the power supply must also be able to produce enough voltage to overcome the wire resistance as it drives current into the coil windings. Electrical power results from the product of current and voltage. The power needs of a high-current–high voltage power supply can be pretty massive, so we grimace at the thought of next month's electric bill. Nevertheless, we hook the magnet up and turn on the current. Rather quickly we discover that resistive heat generated in the wire causes the inner turns to get very hot—hot enough to burn the insulation and melt the copper if we keep the magnet running at very high currents for too long. We can improve matters a fair bit by installing some cooling water pipes alongside the magnet wiring. Unfortunately, the cooling pipes take up some of the space that we really need for magnet wiring, which makes it even more difficult to generate world-class magnetic fields.

Now we try the same construction project using superconducting wire. We choose niobium-titanium alloy—a Type II superconductor—that is flexible, easily formed into wire, and stays superconducting in very large magnetic fields. We wind our coil as before, only this time using niobium-titanium wire instead of copper wire. Operating the magnet in a liquid helium dewar to provide cooling adds a few complications, but not insurmountable ones.

Selecting a power supply to provide electric current becomes surprisingly easier. We still need at least as much current from the power supply as when the windings were copper; field strength still comes from ampere turns. The voltage requirements, however, virtually disappear. In this case, the power supply only needs to provide enough voltage to overcome resistive losses in the normal conductors that feed the current down to the superconducting wiring, but no more. As a result, the power supply eats up only modest amounts of power from the utility lines.

To operate our superconducting magnet, we slowly increase the current traveling through the magnet's coils, and the magnetic field steadily builds. However, this time the growing magnetic field is not accompanied by rapid heating of the conductor—a zero-resistance superconductor. The liquid helium in the dewar continues to silently bubble, boiling away at about the same rate as it did before we turned on the magnet.

The fields generated by superconducting magnets can be truly enormous. Depending on the application, today's superconducting magnets can outdo their copper counterparts in field strength by a factor of three to ten. Such larger magnetic fields immediately translate into larger magnetic forces for electric motor applications. Thus, superconductors give us the option of building either higher horsepower motors or much smaller motors of conventional strength.

Floating Trains

In an increasingly mobile modern society, the problems of transportation concern us more and more. We meet our transportation needs by a variety of means—both with individual vehicles and with public transportation systems. When we choose a mode of transportation, many factors influence our decision. Convenience, cost, and comfort always seem to matter, but when it comes to long-distance trips, nothing has greater importance than speed. Speed made the airlines what they are today. By offering a 10-fold advantage in speed, air transportation steadily supplanted the well-established United States passenger rail transportation system during the middle of this century.

More recently, new technology has allowed railroads to win back a significant piece of the speed advantage enjoyed by the airplane. Especially for regional transportation, where moderate distances are involved, trains are defending and expanding their market niche outside the United States. As an example, the Japanese National Railway has operated its Shinkansen—or bullet train—from Osaka to Tokyo since 1964. Over half of Japan's population lives in the densely populated corridor served by this New Takaido Line. The ability to traverse the entire high-speed rail span in just over 3 hours, while avoiding long waits at airports, has made the bullet train the preferred mode of travel.

Railroad innovators pressing for more speed have done a good job within the constraints of their technology. The high-speed trains of today, including the French TGV and the Japanese bullet train, may represent about the highest level of engineering that can be done with steel wheels on steel track. Designers have made great strides in maximizing the speeds of these trains, but they face some pretty fundamental limits in making further progress.

Train technology stands at an impasse. Although modern high-speed railways have pushed passenger train speeds up to 185 mph (300 kilometers per hour), engineers doubt that much faster service will be possible without a fundamental change in railroad design. Three factors determine the maximum speed of today's passenger rail vehicles: guidance, traction, and power.

The design of railed transportation depends on heavy iron track able to support the tremendous loads of passenger trains. In addition, it depends on secure wheel-to-rail contact to guide the train as it hurdles toward its destination.

Guiding trains on their tracks is not as easy as it sounds. At high speeds, track and wheel imperfections tend to launch the wheels away from the tracks, breaking the solid contact needed for a safe and comfortable ride. Smooth rails are a necessity for high-speed transportation, but we are reaching the limits of manufacturability and maintainability for railroad tracks. To cope with rail and wheel irregularities, heavy suspension systems connect the wheels to the train bodies. Such systems can only function up to a certain speed limit. Eventually, the basic premise of wheel-on-rail train guidance will prevent any further major increases in conventional passenger train speeds.

The second limitation to train speed—traction—also arises directly from the problems of the wheel-on-rail concept. Since the times of Aristotle, we've known that the only way to make something move is to push it. For today's passenger trains, engineers call the *moving force traction*. Traction provides the driving force of steel wheels against steel rails. The principle is no different from the force that propels our automobiles, but a large train requires a much greater tractive force than a car. Unfortunately for high-speed train designers, tractive forces become weaker as rail speeds go up. The same forces that push speeding train wheels away from their tracks also reduce the amount of drive that can be supplied.

Compounding this reduction of traction, there are also drag forces that fight against high-speed motion. *Drag* serves as a catchall term for a variety of aerodynamic and frictional forces that engines must fight in order to keep moving. The bad news for high-speed travel is that aerodynamic drag, the pressure of the wind around the train, increases drastically at high speeds. Again, a fundamental limit blocks major increases in train speeds.

The final factor limiting the ultimate speed of conventional trains involves the delivery of the power required to drive them. We've already seen that large drag forces increase the power demands of high-speed trains. Where do trains get this drive power? For slow-speed trains, diesel engines suffice. The train itself carries all the needed fuel and combustion engine equipment. The engine and fuel are a weighty burden to carry, but at standard speeds, the burden is affordable.

For high-speed passenger trains, the weight penalty for on-board fuel and motors simply becomes too high. If all the equipment and fuel were installed, there would not be enough room for the passengers. To solve this problem, high-speed trains are designed to take electrical power from overhead lines and run on electric motors. The system works quite well at present-day speeds, but keeping sufficient electrical contact at much higher speeds presents engineers with some difficult if not impossible problems.

Therefore, for reasons of guidance, traction, and power collection, conventional railed passenger trains have reached an evolutionary dead end. Minor improvements to rail quality and engine design would increment speeds and performance slightly, but really significant progress such as doubling train speeds requires a real innovation in the basic idea of how trains are designed and operated.

Railroad designers have not given up. Sometimes the best way to combat an insurmountable problem is to go around it, and for passenger trains such a new solution looms on the horizon. Engineers have formed a clear picture of the way around the wheel-on-rail speed limit. The answer involves an entirely new type of railroad track, superconducting magnets, and levitation.

Levitation never ceases to amaze magic show audiences with its conjured defiance of gravity. Good magicians never reveal their secrets, but hidden strings and mirrors usually lurk in the background at these demonstrations. Superconductivity offers quite a different kind of levitation, magnetic levitation, which is neither secret nor confined to a magician's bag of tricks. The magnetic levitation made possible by superconductivity may someday play a major part in our transportation systems.

Picture a levitation force powerful enough to lift an entire train a foot into the air. No wheels touch the tracks and no noisy diesel engine pulls the cars. Instead, this vehicle of the future mysteriously floats on invisible cushions as it silently moves along at 300 miles per hour. Transportation engineers in Japan, West Germany, the United States, and a number of other countries are pursuing such a vision.

There are several ways to turn powerful superconducting magnets into the machines necessary for magnetically levitated (maglev) trains. In theory at least, we could put sets of superconducting magnets on a train and sets of opposite-polarity superconducting magnets in the railroad bed. Repulsion between the two sets of magnets would suspend the train in midair. The drawback of such an approach comes from the cost of all those magnets along the roadbed.

Luckily, two much simpler and much more economical alternative levitation schemes exist. Engineers call the first method *attractive levitation*. Iron is attracted to magnets, and vice versa. By holding a sheet of iron up above a railroad bed, a powerful superconducting magnet aboard a low-slung train can be pulled upwards, lifting the train along with it. The method is simple and inexpensive, but suffers from what engineers call an instability. As the train magnet gets closer to the iron, the levitation force increases and it becomes increasingly difficult to prevent the train from being pulled all the way up into the iron sheet. Practical attractive levitation schemes require a constantly adjusted balance of magnet current and train height to prevent the train from coming to a grinding halt.

The second magnetic levitation alternative is known as *repulsive levitation*. The idea behind repulsive levitation is a little less direct than attractive levitation, but shares the advantage of having a simple passive railroad bed—one without permanent magnets or electromagnets. This type of levitation relies on Lenz's Law of electricity. If we move an ordinary loop of wire next to a large magnet, a current will flow in the loop. The loop attempts to resist any change in the magnetic field that penetrates it. It does so by temporarily producing

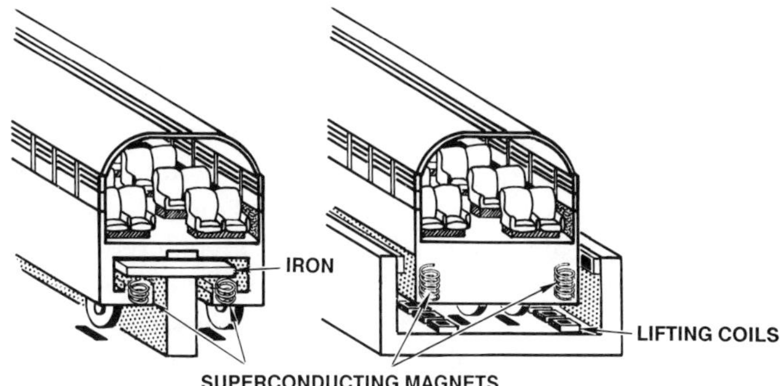

Attractive and repulsive train levitation schemes.

its own magnetic field in opposition to the applied field; this is ordinary diamagnetic current flow.

Engineers exploit this principle in a scheme to levitate the proposed superconducting trains. The idea is to place a series of ordinary conducting wire loops in the path of the train. If a high-speed, magnet-laden train suddenly rushes over such a loop, large currents will spontaneously begin to flow around the loop, generating magnetic fields that oppose the fields emanating from the train. For all intents and purposes, the loop acts like an electromagnet that repels the train's superconducting magnet. The repulsive force levitates the train into the air. Thus with a set of simple wire loops along the railroad bed, we have achieved the same goal that otherwise requires thousands of active magnets to accomplish.

A minor disadvantage of the repulsive track technique is the fleeting nature of the levitation forces. The currents induced in the track's wire loops can only last for a fraction of a second. Resistive losses in the normal conductor rapidly cause the current and its resultant magnetic field to decay away. Fortunately, at high train speeds this doesn't constitute much of a problem. The train magnets pass by faster than the currents disappear. At slower train speeds, on the other hand, the current decay causes reduced lift, or even the total loss of levitation. Engineers solve this problem with a trick borrowed from the airline industry. Just as planes roll along on wheels until they take off, maglev trains can rely on wheels until they reach sufficient speed for levitation. Thus designs for maglev trains include hidden wheels for travel at low speeds near rail stations.

Designers are still debating the relative merits of a variety of magnetic levitation schemes. West German engineers have focused on attractive levitation. Their present designs use conventional magnets, taking advantage of the relatively low magnetic field requirements for attractive levitation. However, small levitation heights and instabilities continue to hamper attractive levitation designs. The research of the Japanese National Railway has centered around the repulsive levitation system, and is fully committed to the use of superconducting magnets. Although faced with larger magnetic field requirements and larger magnetic drag, Japanese engineers have pushed train speeds 100 mph faster than the West Germans. Perhaps the most appealing plan for future systems combines both repulsive and attractive levitation techniques together in a single package. An iron strip provides much of the lift using attractive levitation, but sets of repulsive coils solve the instability problem by applying a correcting force whenever the train deviates from going straight down the railroad path. A complete system includes several superconducting magnets per car along with the liquid helium coolant required to operate the magnets. Designers position the magnets to ensure balanced lift and firm control of sideslip and tilt as well as height control.

Magnetic levitation gives engineers the conceptual breakthrough necessary to seriously contemplate doubling train speeds. Substituting a compliant, invisible magnetic force for the steel wheels of existing trains eliminates the problem of train suspension from the list of factors limiting today's train speeds. The use of special magnetic materials in shaping the fields surrounding train magnets, designers can ensure that trains will remain confined to their guideways. Maglev trains still require special suspension systems for stability and passenger comfort, but the simultaneous horizontal and vertical clamplike containment of magnetic levitation goes a long way toward solving the problems of train guidance.

The second and third limitations to train speed—traction and power delivery—can also be overcome with the help of superconductivity. At first glance, we might expect magnetic levitation to compound these problems. Aerodynamic drag still presents an unavoidable power drain on any train system, but now we can't even supply the additional drive power needed through the train's wheels. To make matters worse still, magnetic levitation adds an additional electrical drag force, one that acts just like aerodynamic friction. Because of these two effects, a high-speed maglev train will require 10,000 horsepower or more in order to maintain its velocity.

Even forgetting about fuel requirements or alternate power sources, the weight requirements of such a 10,000-horsepower engine would greatly tax the carrying capacity of any magnetically levitated vehicle. The problems of both driving force and power delivery require a solution.

The superconducting solution to the drive problem eliminates the need for transferring any power to train engines. In fact, superconducting trains can travel without any on-board engines at all. Instead, a series of wires set in the railroad bed provide all the necessary power. Passing current through these wires generates a magnetic field that attracts the train magnets and thus pulls the train forward as it approaches. Reversing the field as the train passes overhead allows the same track magnets to repel the train magnets and thus push the train away as it leaves. Engineers call such a system a *linear synchronous motor*. The system is synchronous because it requires the track currents to be synchronized to the train motion, and it is a motor because its fundamental operating principles are the same as an ordinary revolving motor, except that its components are stretched along a straight line.

The timing of track currents for the linear synchronous motor turns out to be easier than it sounds. Engineers energize the track with alternating current. Controlling the relative timing in a series of three different track current lines produces a magnetic wave that travels down the railroad bed at a fixed speed. As long as the acceleration is not too drastic, the train will automatically lock onto this traveling magnetic wave, just like a surfer atop an ocean wave.

Linear synchronous motors will bring an end to runaway or overspeed trains. System engineers can precisely control the speed of levitated trains by setting the frequency, and hence speed, of the propulsive magnetic wave along the guideway. The job of the on-board railroad engineer reduces to telegraphically asking track controllers for speed changes.

With the basic limitations of guidance, traction, and power delivery overcome through superconducting levitation and the linear synchronous motor, superconductivity clears the way for truly innovative train systems. The theoretical barriers have come down. Engineers have now begun the long process of proving the technology in practice to test its practical feasibility.

After drawing board sketches and calculations are completed, a

Japanese superconducting Maglev train. (Photograph by Kaku Kurita; Gamma/Liason)

project such as the maglev train proceeds to scale models. The most advanced model project to date was completed by Japan National Railway in the 1970s.

Engineers from Japanese National Railway constructed a series of prototype experiments, beginning with a scale model tested in 1972. The test vehicle, called the ML-100 (for Magnetic Levitation), weighed 3.5 tons and was levitated by four superconducting magnets cooled by on-board liquid helium. The prototype could seat two, traveled on a 480-meter track, and ran at relatively modest speeds up to 35 mph. Since then, the Japanese have tested an unmanned test vehicle at speeds up to 320 mph and a manned, three-passenger train up to 250 mph, both of which far exceed the maximum speeds possible with conventional railed transportation.

Based on their data, the Japanese workers envision a full-scale train with sixteen 30-ton cars, each with eight levitation magnets. The takeoff speed would be 60 mph for levitation, with a 340-mph maximum speed. A total of 300,000 electrical horsepower (100 megawatts) of energy gets supplied by the linear synchronous motor track to propel the vehicles. This constitutes quite a lot of power but, on the other hand, only track sections that are actually near trains must be energized.

Research on magnetic levitation transportation systems peaked in the early 1970s and then entered a period of engineering dormancy. The decreasing price of gasoline and the increased efficiency of airplanes have limited the desire for expensive new construction projects on unproven large-scale systems. The recent advances in new superconducting materials, however, have begun to reawaken interest in magnetic levitation.

Superconducting magnetic levitation has the potential to reverse the decline of U.S. passenger rail transportation. By whittling away the velocity differential between trains and planes, superconductivity may help tip the balance for time-conscious travelers. For short- or moderate-distance trips, such as between Washington, D.C., and New York, time saved in travel to terminals, boarding, and runway delays could easily make maglev train service the faster way to go. If so, air transportation could some day only make sense for long-haul, transcontinental routes.

Implementation of superconducting magnetic levitation for high-speed trains will be an expensive proposition. Accurate predictions of right-of-way, power system, and vehicle costs are hard to come by, but estimates of around $50 million per mile appear to be reasonable.

Surprisingly high land costs act in the favor of rail transportation over expanded air service. Rail transportation makes very efficient use of real estate. In fact, the land requirements for Japan's entire bullet train system are smaller than what was needed for Narita airport, a single facility outside of Tokyo. A superconducting train line should not require much more right-of-way than the more conventional bullet train.

A number of routes appear to be particularly attractive sites for superconducting maglev train systems. Japan has a good chance of leading the way with the first such system. Both political and demographic factors act in Japan's favor. The bullet train, operating on the Tokyo-to-Osaka route, has been an unqualified success. Over half the population of Japan lives in the Shinkansen service area. A maglev train system that cuts the longest transit times to under 2 hours would be a very popular choice.

In the United States, engineers eye the northeast corridor, stretching from Washington, D.C., to Boston, as a candidate site for maglev trains. This region is densely populated and heavily traveled. A well-established network of air carriers operating in the area will undoubtedly discourage any substantial transition back to rail transport.

Even though there are no full-scale maglev train systems under development, scientists and engineers are already looking ahead to the eventual practical problems of such systems. Tests and proposals to date have mainly centered around the performance of well-proven niobium-titanium wire technology operated in a liquid helium bath. High-temperature superconductors are causing engineers to rethink many levitation designs, both in train magnets and in the railroad bed. Levitation field strength, refrigeration power, and magnet weight all become more favorable with liquid-nitrogen-cooled superconducting technology.

Other long-range plans center around even further increases in speeds. Successfully competing with air travel will require maglev trains to attain 300 mph, performance that stands as a primary objective. Although the everyday use of such trains may still be decades away, engineers and scientists are already tossing around ideas for even more advanced systems for maglev transportation. One suggestion calls for the construction of evacuated tunnels for magnetically levitated trains. Such trains would require pressurized passenger compartments, much like on airliners. By removing the air from the transport tubes, the effects of aerodynamic drag disappear. As electrical drag decreases at high speeds as well, supersonic trains may even be possible.

Superconductors at Sea

Even though land transportation looks like a promising field for superconductor applications, superconductivity might not make its first appearance in a practical transportation system on dry land at all. Instead, it may prove itself first on the high seas. The United States Navy has been actively investigating the use of superconductors in a whole new way of propelling their ships. Today, all propulsion systems aboard naval vessels are quite similar. Some sort of power plant—either nuclear, diesel, or steam engine—provides the raw power for ship propulsion. In all cases, the power is transmitted to the ship's propellers via a series of turbines and gears. Conventional propulsion units impose severe constraints on ship designers. Since propellers first replaced paddle wheels, all ships have shared the same pattern. For efficiency and mechanical strength, the generating plant, gears, propeller shaft, and propellers must lie along a straight line. This arrangement inevitably takes up significant amounts of valuable shipboard space. Like it or not, all other equipment and cargo must be fit around a ship's engine plant.

An alternative exists for shipboard propulsion systems and it consists of linking the power plant and propeller electrically, rather than mechanically. Electric propulsion replaces the mechanical gearing and heavy drive shaft with a generator and a motor. The power plant turbine spins a generator and makes electricity. Wires then carry the power to electric motors that couple directly to the propellers. It may sound cumbersome to start with mechanical energy, convert to electrical energy, and then convert right back to mechanical motion, but the scheme offers some important rewards. For one thing, changing the speed and even the direction of the propeller can be done electrically, eliminating the need for heavy gearing equipment.

Electric propulsion leads to innovations in how we design our ships. If we incorporate the additional size and weight advantages of superconducting components, electric motors can free ship designers to plan novel and improved ship layouts. The generator and motor do not need to be solidly joined in a rigid line. They don't even have to be anywhere near each other. Designers can place generators anywhere in the ship, being free to cater to considerations of weight balance, survivability, and hull profile.

The freedom to place the drive mechanism where it can do the most good rather than where mechanical constraints forces it to be excites the imagination of ship designers. Small but powerful electric motors can connect directly to propeller blades. Designers call such a unit a *pod*. Because the pod connects to the power plant with a simple, bendable power cable instead of a heavy mechanical shaft, pods can be placed just about anyplace. The old straight-line propeller shaft limitation disappears. Instead of the standard propeller mounting directly behind the ship, pods with propellers can be placed directly underneath the ship for better efficiency. Pods might even be able to pivot for maneuverability. Designers could optimize pod placement for low noise, making ships harder to detect from their propeller sounds.

Superconducting generators and motors can handle the enormous power requirements (something like 40,000 horsepower for a typical destroyer) required for electrical ship propulsion. Furthermore, superconducting systems are lighter and more compact than their conventional conductor counterparts. Fully aware of these natural advantages of superconductivity, the United States Navy has pursued superconducting shipboard propulsion since the late 1960s. In order to master the many complexities of superconducting propulsion, Naval researchers embarked on a carefully planned series of demonstration projects, involving larger and larger superconducting units, and designed to put superconducting propulsion to the test.

To check out drawing-board concepts in realistic situations, the Navy dedicated a 65-foot "test craft," christened the Jupiter II, as a test vehicle for superconductive propulsion. Experiments performed by the David Taylor Naval Ship Research and Development Center in Annapolis have succeeded in demonstrating the feasibility of superconductive drive. The first set of experiments, carried out in the 1970s, culminated in a 400-horsepower superconducting generator matched to an equally powerful superconducting motor. Although not tremendously powerful by naval ship standards, the propulsion system proved that an all-superconducting electrical propulsion system could successfully handle the needs of an ocean-going vessel.

A second phase of the Naval program produced a larger superconducting generator and motor set. A 3,000-horsepower system was installed on the Jupiter II in 1983. The unit functioned successfully in a number of sea trials and made a convincing case for the feasibility of superconducting propulsion.

Since the early 1980s, progress has slowed somewhat. The current status of superconducting electric ship propulsion remains in some doubt. Electric propulsion still looks like a very good idea. The debate centers around whether to use a superconducting approach or rely on more conventional, copper-wound generators and motors. Both methods are practical. Faced with limited budgets, Naval planners decided to concentrate on conventional conductors such as copper for their working generators and motors. The United States Congress has authorized funding for electric ship propulsion for the Navy, and the Navy is looking for a contractor able to supply the necessary drive units.

Superconductivity has not yet been eliminated from serious consideration. In response to the discovery of high-temperature superconductivity, Congress has earmarked a portion of the appropriated funds for the Navy to investigate use of new superconducting mate-

The Jupiter II, a ship propelled by superconducting motors. (Courtesy David Taylor Naval Ship R&D Center)

rials. The eventual design of electric propulsion systems may well hinge on developments in these new high-temperature superconductors.

If high-temperature superconductors reach the point of technological viability, the scales may tip in favor of superconducting electric propulsion systems. Liquid-helium-cooled superconducting systems appear capable of doing the job, but have thus far lost out to conventional electrical systems because of the perceived technical risk associated with liquid-helium cooling. Liquid-nitrogen-cooled superconductors may change the situation significantly. For one thing, the Navy would prefer liquid nitrogen over liquid helium for logistical reasons. While helium requires transportation from natural gas well heads in Louisiana and Texas, nitrogen can literally be distilled from thin air anywhere in the world. Furthermore, there are far fewer problems associated with the handling of liquid nitrogen and liquid-nitrogen-cooled equipment.

In future ocean-going vessels, the engine room may be only one of several places where superconductors can be found. A technology based on liquid-nitrogen-cooled superconductors would substantially reduce the costs and overhead burden associated with refrigeration plants. As a result, naval designers can contemplate using the advantages of superconducting generators and motors in smaller scale applications, such as on-board ship power generation. Superconducting electric propulsion may also become practical for smaller ships, such as minesweepers. Many uncertainties remain, and naval designers watch the developments of superconducting materials, wires, and machinery with great interest.

Magnetic train levitation and large ship propulsion appear to be natural candidates for the application of large magnetic fields generated with the help of superconductivity. Serious investigations were underway decades before the discovery of ceramic superconductors, and even more critical thinking goes on today.

Superconductors on the Road

Even our favorite mode of transportation, the automobile, may benefit from applications of superconductivity if the new high-temperature superconductors can be successfully developed. Some of the proposals for superconductivity in highway vehicles will require major advances in technology, but are interesting to consider anyway. Using magnetic storage to replace batteries in an electric car has received some serious consideration. As we saw when we discussed electric power applications, superconducting magnets can efficiently store energy for long periods with virtually no loss. A superconducting coil could serve as a magnetic battery, supplying and storing electrical energy in automobiles. Unfortunately, according to current design ideas, the magnetic field strengths, the mechanical strength of the magnet windings, and the physical volume of the magnet coils required for such applications appear to make such systems rather impractical. New materials and techniques will be required to make such techniques more attractive.

We would certainly imagine that dozens of small-scale applications for personal automobiles would appear if room-temperature superconductors became available. Today's automobile contains many electric motors. Starter motors, alternators, windshield wipers, seat adjustments, power locks, and power windows all use electric motors. A superconducting technology that could shrink the size and weight of any of these motors could quickly pay for itself in increased fuel economy. Similarly, a room-temperature superconducting wiring harness, one that replaces heavy copper wiring bundles with thread-like superconductors, would also find a place in the car of the future.

Perhaps the most intriguing suggestion calls for levitated automobiles, akin to the already demonstrated maglev trains. Just before entering special magnetic super-superhighways, drivers would energize superconducting magnet coils near the wheels. At the entry ramp, road magnets would take over control of the car, quickly accelerating the vehicle up to perhaps 150 mph. Even at these speeds, bumper-to-bumper traffic would be quite safe, as the levitation system tightly holds each car in a vicelike magnetic grip. Of course, such a highway system would never see a speeding ticket issued, or a flat tire, although safeguards against magnet failure would be essential.

Major changes in transportation systems are generally slow to happen. Our interest in such developments waxes and wanes with the ups and downs of the oil industry. This state of affairs makes it difficult to predict the timetable for the introduction of futuristic transportation technologies. Nonetheless, the chances are quite good that the twenty-first century will see an increasing use of superconducting magnets and motors in ground and sea transportation.

CHAPTER 9

Fast, Quiet, and Precise
Superconductive Electronics

In the 1980s, when we hear the term "high technology," we immediately think of electronics. In a world with satellite television, compact disc players, electronic appointment-calendar wristwatches, and powerful desktop computers, we are surrounded with seeming miracles of electronic technology. We have come to believe that today's microchip designers can perform any magical feat we can dream up, and at affordable prices, no less. Customers for really high-performance electronics have the same thought in the backs of their minds: modern electronics can do anything.

Reality has a way of catching up with our dreams, and for high-performance electronics, it is catching up fast. What people want out of electronics for even the near future can be expressed as a series of simple demands: it has to be faster, smaller, quieter, higher frequency, more sensitive, more energy efficient, and, in short, better than ever before. The electronics industry has lived under the gun of these same demands for two decades, so none of this is news. But a specter that looms on the horizon turns out to be the physical limitations of conventional electronics. In one area after another, electronics will finally reach limits that cannot be exceeded because of fundamental laws of nature and the intrinsic properties of materials. The only solutions will be alternative electronics technologies. Superconductive electronics represents one such technology that holds tremendous promise.

At least with the current crop of superconductors, superconductive electronics is not likely to find its way into wristwatches and stereo systems. It is, on the other hand, very likely to show up whenever the heavy artillery gets called out for electronics applications. The fastest, quietest, lowest power, highest frequency electronic circuits in the world are all superconducting circuits, and more people want and need them every day. Viewed from a historical perspective, the same key advantages that transistors held over vacuum tubes many years ago, superconductive electronics now holds over transistors. Before we examine some of the places in which superconductive electronics will make its mark, we consider the origins of some of its intrinsic advantages.

The intrinsic speed of superconductive electronics comes from the rapid switching ability of Josephson junctions. Applying current to Josephson junctions can make them switch from the zero-resistance state to the finite-voltage state in trillionths of a second (picoseconds). Circuits made from such junctions can therefore control electronic signals in similar time frames. The most advanced semiconductor devices can compete at nearly a comparable level in this picosecond time scale, but only at the cost of much higher operating power. For reasons we will soon see, high-speed electronics demands miniaturization but high power and miniaturization cannot coexist.

When we talked about Josephson junctions, we learned that they pass current either at millivolt (thousandth of a volt) levels or with no voltage at all. In contrast, the characteristic signal levels on transistors are a few volts. Because of this 1000:1 voltage ratio, transistor circuits require thousands of times more power to operate than Josephson circuits. In very densely packed electronic circuits, all this extra power not only results in inflated power costs, it can also literally burn up the circuit. A circuit's ability to shed waste heat depends on how much the heat is concentrated. Whereas large and sparsely populated circuits can easily shed excess heat, a densely packed circuit full of power-hungry semiconducting devices becomes vulnerable to self-destruction.

Josephson junctions also distinguish themselves by their low noise. The term *noise* deserves some explanation. Electronic noise isn't the noise of a noisy party, but more closely resembles the indistinct hissing we hear on the radio and on the telephone, or the snow that we see on a television screen. All electronic devices add noise to signals as a result of fundamental processes that take place on the atomic level. The same collisions that lead to electrical resistance in a lattice will also broadcast small noise signals in electronic circuits. Such noise is truly ubiquitous. Engineers study noise very carefully because the ultimate sensitivity of any circuit depends on the amount of noise present—noise that can obscure the faintest signals.

The collisions in a lattice that create noise are fueled by temperature. Colder electronic circuits are intrinsically quieter electronic circuits. With the benefits of zero resistance and liquid-helium temperature operation, superconductive electronic circuits operate with almost undetectably low noise levels. With such nearly noiseless circuits, scientists are able to undertake experimental investigations that would have been unimaginable without superconductivity. Thus, they can measure currents within the brain, probe seismic forces deep within the earth, and listen to the dying echoes from the cosmic big bang—all by using superconductive electronics. As we will see, superconductivity has a part to play throughout the world of electronics.

Modern electronics has now split into two distinct types: digital and analog. Superconductive electronics can meet the demands of the most challenging applications of both digital and analog electronic technology. In the digital world, the driving force is speed. In the analog world, the quest is for ever-higher-frequency operation. To see how superconductivity fits in, we need to understand a little about analog and digital electronics. The distinction comes from the way in which circuits handle information.

Electronic circuits transmit information in the form of signals that change with time. Analog electronics makes use of signals that mimic the original information. For example, the sounds that emanate from a violin can create electronic signals in an analog device called a microphone. These signals are made stronger by another piece of analog equipment (an amplifier) and then can produce mechanical motion in another analog device: a loudspeaker. At every step of the process, the signal is a replica of the original sound wave. When the sound gets louder, the electrical signal gets larger; when the pitch of the violin goes up, the frequency of the electronic signal increases in kind.

Digital electronics handles signals in an entirely different way. The sound wave from a violin does not become an electronic signal that changes in time, it becomes a list of numbers. The numbers represent the amplitude or strength of the wave measured at successive instants. Such a digital representation is analogous to capturing motion on movie film. A sequence of still photographs reveals all the motion in the original scene when projected at the proper speed. Similarly, a sequence of amplitudes reveals an original waveform when played back at the proper sampling rate. To achieve complete fidelity to the original signal, we would have to record amplitudes at every single instant in time in order not to miss a moment of the signal. But doing that corresponds to analog signal processing. The key to digital signal processing is that we don't need to measure amplitudes on a continuous basis to accurately capture a changing signal. We know that movie film accurately records motion as long as enough still frames are recorded. If too few pictures are taken, filmed motion becomes jerky, but if the camera snaps pictures fast enough, the movie will look like real motion. Similarly, digitization of signals has minimum speed requirements. A mathematical rule called Nyquist's Theorem sets the requirements for digital signal processing. As long as enough amplitudes are recorded, a signal can be represented with a list of numbers that is much more compact than any replica of the original signal.

Superconducting Supercomputers

Electronic circuits are well suited to handling numbers instead of waveforms because numbers can be represented in a particularly simple fashion. Digital electronics uses the binary representation for numbers (base-2 arithmetic), a system in which any number can be written as a series of zeros and ones. Electronic circuits have a built-in way to represent zeros and ones: they can be on or off. This gives digital circuits a great advantage over analog circuits in terms of simplicity and integrity of information. Whereas analog circuits can fall prey to noise and other problems that can obscure subtle features of their signals, digital circuits only need to be able to tell the difference between on and off.

For most of the history of electronics, all circuits were analog circuits. Radio, television, stereo, and telephones were all developed as analog circuits. Digital electronics made its first appearance in the form of computers, and only recently have other familiar electronic circuits become digital circuits as well. But while most of electronic technology began with analog circuitry, superconductive electronics got its start with digital applications.

Superconductive electronics has promised ultimate performance for a number of technologies, but none has attracted more attention than high-performance computing. No other field has tougher requirements. Today's high-speed computers must make literally billions of electronic decisions every second, and make them without a single error. Tomorrow's computers will have to be faster still in order to handle the increased flow of information in our electronic society and to solve the ever more complex problems being posed by scientists and engineers.

While the demands are high, so are the payoffs. American computer manufacturers are approaching the $100-billion mark in annual sales. The reason for the industry's growth is simple: both science and industry have found that fast computers more than pay for themselves in cost and labor savings. In the constant evolution of computer designs to improve performance, superconductive electronics may be the technological leap needed for the next generation of computers.

The past few decades have seen ever-faster computers. Even the personal computers of today run at speeds beyond the limits of the largest computers of the 1960s. Today's supercomputers can solve immensely complicated problems with amazing speed. Just who is it that needs computers faster than today's most powerful electronic brains? We find that scientists and engineers from a number of varied disciplines face problems that, although theoretically solvable, are just too big for even today's fastest computers to solve in a practical amount of time.

As an example of high-speed computer applications, consider the problem of forecasting the weather. Meteorologists could make use of almost unlimited amounts of computer power. Starting with careful measurements of the day's weather, and with the aid of a high-speed computer, forecasters could build an electronic model of the earth's atmosphere and tell us what weather we could really expect tomorrow. For such forecasting to work, a computer must be fast enough to factor in many atmospheric variables—like atmospheric pressure, temperature, and humidity—on a point-by-point basis over large areas of the earth and on an instant-by-instant basis in time. Existing computers can't possibly manipulate this quantity of data fast enough to predict the weather very far in advance. A big enough and fast enough computer could enable such detailed and accurate modeling to be performed for time scales as long as a week, or perhaps even a month.

Airplane designers want faster computers, too. No simple equation tells an airplane manufacturer how to shape the most fuel-efficient or fastest airplane wing. Often designers rely on the old procedures of building detailed models to be used in wind tunnel tests that measure wind drag and characterize air flow patterns. A fast computer could tell how any given design responds to the aircraft environment very exactly and in great detail. With a few strokes on the computer keyboard, engineers could test new wing designs without worrying about the effects of wind-tunnel imperfections or model irregularities.

The large payoffs for computer speed have prompted engineers to employ a succession of different technologies in order to achieve ever-increasing computer performance. Engineers have built computers using mechanical, electronic, and even pneumatic technologies. But despite their outward differences, all digital computers are identical at heart.

Keeping track of lots of numbers stands as the obvious first requirement for any computer technology, old or new. The abacus records numbers using the positions of beads along a set of rails. Builders of mechanical adding machines accomplish the same job with interconnected series of gears and levers. The first big jump in speed came with electronic computers, in which numbers are handled as electric voltages. The large electronic computers created during World War II relied on vacuum tubes—the same glowing glass bulbs that used to fill our radios and televisions. Later, engineers discovered that transistors could accomplish the same electronic functions as vacuum tubes by electrically controlling the flow of charge in semiconducting materials.

Transistors replaced vacuum tubes because they are smaller, faster, and consume less power. As the 1960s ended, the electronics industry made another leap forward in computers with the introduction of integrated circuits. By placing many transistors on a single tiny chip of semiconducting material, computer makers could build circuits that were faster and smaller than ever before.

For 20 years—quite a long time for a rapidly moving technology—semiconductor integrated circuits have held the high-speed computing records. Circuit engineers have steadily pushed their designs to higher complexity, density, speed, and computational power. Powerful market forces, daring venture capitalists, brilliant innovators, and inventive computer "architects" have joined forces to create today's supercomputers. But even as incredibly powerful computing machines like the Cray II supercomputer have become available, the need for even higher performance computers continues to grow.

To meet this need, engineers have to find new ways to build computers. For years, they have been evaluating a number of competing technologies for the computers of the future. Josephson technology enters the competition with an impressive list of natural abilities and some clear-cut advantages over any semiconductor technology.

The fundamental functioning of a digital computer involves some electronic means of distinguishing between on and off, or one and zero—the basic language of digital electronics. All the information contained in a computer consists of ones and zeros; things only become complicated because there are huge numbers of these binary digits (bits) and the computer has to work with them at blinding speed. In semiconductor technology, engineers developed semiconducting devices whose resistance would alternate between two values whenever a voltage was applied, thus establishing the two states equivalent to "0" and "1." Superconductors, on the other hand, can switch between a resistive state and their characteristic zero-resistance state under the influence of temperature, current, or magnetic fields. The two resistance states of superconducting materials provide a natural mechanism for the manipulation of digital information. In order to build a practical digital switch from superconductors, device designers just needed a way to rapidly control the superconductivity in a small piece of metal. The first such devices were called *cryotrons*, which were literally strips of superconductor that could be driven into the resistive state under the influence of current. Engineers investigated a whole family of devices related to the cryotron, including ones that used heat, magnetism, current, and even injected electrons to turn superconductivity on and off. Eventually, they found that all these methods were lacking in speed or practicality. The discovery of the Josephson junction in the 1960s provided the solution to the problem of finding a practical superconducting digital device.

The building-block function we need for computer circuits consists of a way to unambiguously store a one or a zero in a circuit element. The Josephson junction provides a wonderful way to do the job. We refer to the same current–voltage characteristic of a Josephson junction that we looked at in Chapter 3. This curve tells us

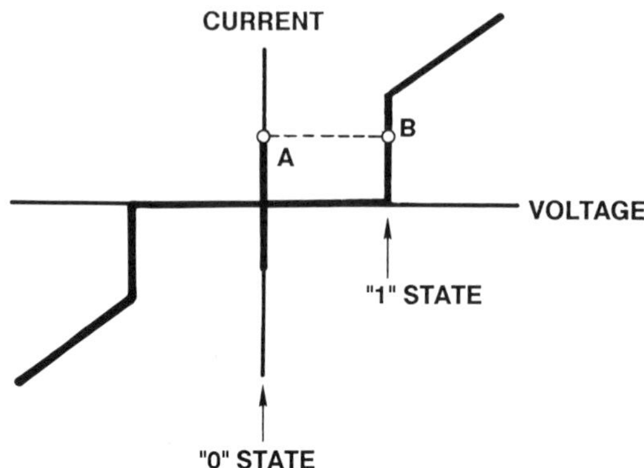

"Zero" and "one" states of a Josephson junction.

what voltage will appear across the junction when a given amount of current flows. Suppose we apply just enough current so that the junction functions at point A on the diagram. Then the voltage across the junction will be zero. We thus define a "zero" state for our computer circuit. Now suppose we want to change the state of the circuit to a "one". We simply increase the current a small amount and the junction will switch to point B on the diagram. Now the voltage will be equal to the energy gap value for the superconductor, typically a few millivolts. We can easily distinguish a few millivolts from no voltage at all, so we have a well-defined "one" state for our circuit. The device turns out to be very stable as well; in order to return to the "zero" state, we must turn off the current entirely to restore the zero voltage current.

Thus Josephson junctions provide a simple way of defining the two distinct states that are the essence of digital electronics. However, there are already many ways to accomplish the same thing without superconductors. The reason that Josephson junctions stand out from the crowd is that they do the job incredibly quickly and use very little power in the process. In terms of speed, the jump from the zero-voltage state to the gap-voltage state in a Josephson junction can occur in as little as a few picoseconds (trillionths of a second). In that switching time, light cannot even travel the distance between these two periods.. A computer writing names at one every two picoseconds could list the entire world population in a hundredth of a second. The computers of tomorrow will require switches of this kind of of speed. So we see that superconductivity can meet the switching-speed requirements of a state-of-the-art computer. But switching speed alone cannot produce the world's fastest computer.

Whether a Josephson computer would really give super-high performance depends on some rather mundane but highly important side issues like wiring and system integration. In the computers of the future, limiting speeds will not be set by the speed of the logic devices, but instead by the quality and length of the wires that connect the subcircuits within the computer.

Suppose you accomplished the impossible and built an electronic switch that could be turned on or off in zero elapsed time. Could you then build an infinitely fast computer? Alas, the answer is no. After doing each calculation step, the computer would still have to wait long enough to make sure that the output data had reached a safe storage place in memory, and then wait again to be certain that the next instruction had enough time to travel from the memory back to the computer. Operation of super-fast computers becomes a waiting game as digital signals traverse the electronic real estate within the computer.

The electronic information travels at nearly the speed of light, the fastest attainable speed in the universe. But for some computer applications, even that speed can seem annoyingly slow. Today's fastest computers function on timescales measured in nanoseconds (billionths of a second). A nanosecond corresponds to only enough time to allow light to travel a foot. Because of this, supercomputers face a natural size limit. Any computer that breaks the nanosecond barrier will have to be no larger than a teacup in order to avoid being slowed down by the signals traveling within it.

Advances in microscopic patterning techniques have miniaturized transistors and integrated circuits, but as we already pointed out, simple heat generation limits the ultimate size of such devices. Semiconductor chips must always be made large enough to allow the heat they generate to escape into the computer environment. The fastest semiconductor circuits burn the greatest amount of power. Even with heroic cooling techniques, a sub-nanosecond semiconductor computer may still cook itself to death.

Superconductivity holds the promise of computers smaller than can be imagined with today's transistorized technology. The low power dissipation of Josephson junctions allows tighter packing of the logic circuits, producing physically smaller computers. Naturally enough, the tinier the computer, the less time it takes for signals to travel around. The teacup-sized computer appears to be a goal within reach of superconductive electronics technology.

Superconductivity has a second, more subtle advantage over conventional integrated circuit technology that doesn't depend on the properties of Josephson junctions at all. Instead, it relates to the high-speed electronic pathways that carry information around computer circuits. One can visualize the flight of electronic information within a computer as flowing along a miniature network of telephone lines. Each wire within the computer carries a one-way digital conversation from one logic element to the next. For the computer to work, these electronic messages must pass quickly and clearly around the computer circuits. Ordinary computer wiring has resistance. Wire resistance robs digital signals of some of their strength, making it harder to tell the difference between "1" and "0" signals. Even worse, resistive wire twists and distorts high-speed signals, garbling digital information to the point of incomprehensibility. The effect, called *dispersion*, becomes especially severe for the fastest changing signals. The faster we push semiconductor circuit elements, the more dispersion becomes a problem. Superconducting transmission lines, on the other hand, have negligible dispersion at typical signal frequencies.

Dispersion limits the speed of super-fast semiconducting computers and conventional conductors cannot solve the problem. As a result, superconductivity may first appear in computers in the form of simple wiring. Superconducting wiring has additional advantages as well. Superconducting wire of a given dimension can carry a thousand times more current than the same-sized copper wire. Conversely, for fixed current levels, superconducting wire can be made much smaller, liberating precious chip real estate for other purposes.

Computers also have to have memory; information must be stored in the machine. The superconducting solution to the memory problem comes from flux quantization in superconducting rings. As we recall, applying a magnetic field to a superconducting ring results in a current flowing around the ring that assures that the magnetic flux within the ring comes out to be an allowed value. To store information in a superconducting computer, tiny currents generate magnetic fields which in turn induce currents in superconducting rings. The zero-resistance currents flow indefinitely, storing a specific current value. The ones and zeros of computer memory correspond to currents that flow clockwise or counterclockwise around the loop.

Superconducting memory is quite simple in principle, but has proven in practice to be difficult to build.

Scientists have been working on all the required subsystems for a Josephson computer for some time now. The basic strategies of the superconducting computer, namely the use of Josephson junctions as logic elements and superconducting rings as memory, were recognized in the mid-1960s. At that time, several research laboratories began to experiment with Josephson logic. A concerted effort to develop a commercial Josephson computer began, with the most serious effort, at least in terms of committed manpower and capital, coming from IBM.

Starting in 1968, the management of IBM research committed substantial resources to developing the technology required for building a supercomputer based on superconductivity. The goal was nothing less than the birth of the next generation of computing machine. To back their efforts, IBM spent well over $100 million over a period of 15 years.

Almost overnight the sociology of superconductivity research changed. Superconductivity, long studied only in small university research laboratories, was thrust into the corporate environment. Systems specialists partitioned their computer requirements according to discipline. Integrated circuit designers, materials scientists, low-temperature physicists, and cryogenic engineers assembled into teams and departments to tackle specific issues. The superconducting computer suddenly faced the analytical scrutiny of a large, well-financed, organized community.

Constructing a computer from the superconducting technology available in the 1960s might be compared to mass-producing automobiles starting from Soapbox Derby expertise. The promise was there, but problems of large-scale fabrication challenged the resourcefulness of the computer designers. As we saw earlier, in the 1960s, making even one reliable Josephson junction proved to be a tremendous technological challenge. Building even the simplest computing circuit, one containing perhaps a dozen junctions, required seemingly endless fabrication attempts. Compared to building test circuits, the creation of a complete superconducting computer would stretch fabrication requirements to new dimensions. While test circuits consisted of a few dozen junctions, thousands of junctions were needed for even the simplest memory chip and all those junctions needed to be able to survive hundreds of cycles between room temperature and liquid-helium temperature. Reproducibility and reliability arose as the primary concerns. The circuit engineers' designs called for adjacent junctions to have nearly identical performance. To make things even more difficult, junctions on different chips, possibly manufactured months apart, also had to be identical. Fabrication techniques like these are taken for granted in automobiles and stereos, but producing superconducting circuits was a technology that had not yet risen above the level of a tinker's craft.

As we saw earlier, IBM applied the tools and techniques of semiconducting integrated circuit technology to the problem of Josephson circuits and made enormous strides. By the mid-1970s, IBM's demonstration circuits began appearing in the scientific literature. A 1975 paper reported a Josephson multiplier circuit that could multiply together numbers as high as three. This sounds like a preschool task, but constituted a major milestone in the development of superconducting integrated circuits. The chip was relatively slow by today's standards, but no semiconducting test equipment could keep up with it; the Josephson multiplier had only mastered a very simple multiplication table, but it was fast! Of equal significance was the fact that 50 Josephson logic subcircuits went into the operating design—a level of complexity that was unheard of at the time. Modest as this first chip was, Josephson logic had reached the gateway of integrated circuit technology.

In addition to the chips themselves, the computer's electronic environment began to take shape. Superconducting chips had to be bonded onto cards and the cards fitted together into frames. Engineers had to find ways of getting high-speed logic signals from one side of the computer to the other. Designers devised special interconnecting boards with all superconducting wiring and used balls of mercury to establish wiring contacts.

As the Josephson project progressed to the point of medium-scale integrated circuit fabrication, the complexity of the tasks increased tremendously. The years since the beginning of the project saw huge advances in semiconductor computer complexity, and Josephson technology was called on to match that progress to prove its viability.

IBM's leadership in developing a Josephson computer technology came to an abrupt end in 1983. Management's continuing technical assessment of the program had begun to cast doubts upon its viability. It saw a Josephson technology that had advanced far be-

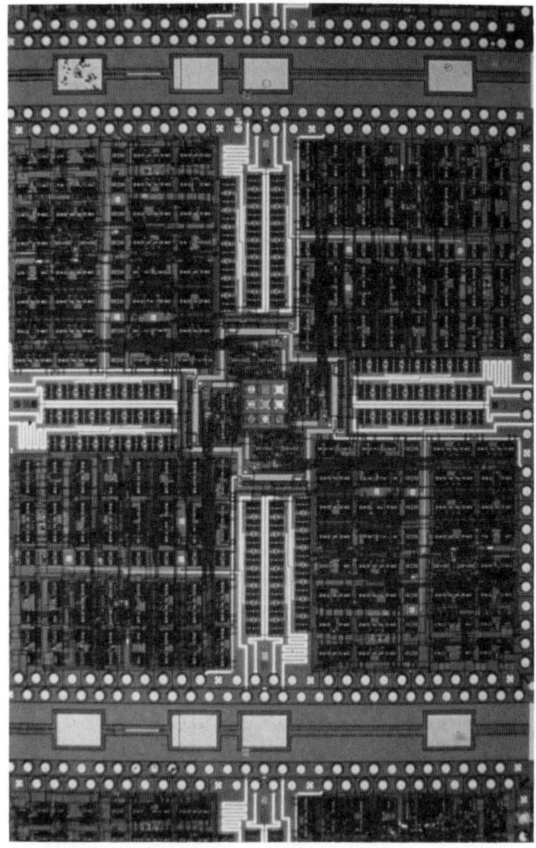

Superconducting circuitry from the IBM computer project. (Courtesy IBM)

yond its infancy of two decades earlier. But serious technical challenges remained in both electronics and materials. There were much-improved transistors made from silicon and gallium arsenide with speeds projected to be close to the domain once exclusively belonging to superconductivity. IBM management saw the practical problems of building superconducting memory circuits as requiring many more years for acceptable solutions. So, for a combination of reasons, IBM decided to shift its emphasis to a more balanced array of high-performance technologies and reduce the activities in superconducting technology to a small, research-oriented effort.

The widely publicized exit of a computer superpower like IBM from the digital superconductivity field coupled with reductions in superconductive electronics programs at several other labs had a chilling effect on research throughout the superconductive electronics research community. Suddenly the gamble of research and development dollars on a promising but unproven new technology appeared to have become a much higher risk. But reports of the death of superconductive electronics were, in the words of Mark Twain, greatly exaggerated. The IBM program had set new standards for superconductive electronics performance and had demonstrated that complicated circuits could be built with Josephson junctions. In the United States, the time had come for scientists to set their sights on more attainable goals in Josephson technology.

But in the meantime, IBM did not in fact deal a finishing blow to efforts to build the first Josephson superconducting computer. In Japan, the Ministry of International Trade and Industry (MITI) united the forces of industry and academia in a nationwide, coordinated attack on the supercomputer problem. Japanese researchers are producing superconducting circuits that reach new levels of complexity and speed. A few years after the end of the IBM program, researchers at Hitachi reported success in building a complete sub-nanosecond Josephson data processor. While data processors are only medium-sized building blocks in a true supercomputer, the Hitachi processor features an impressive range of subcircuits. Incorporated on a chip only a tenth of an inch square, the circuit contains special power regulators to provide both operating current and timing signals to the logic circuits, a decoder to interpret incoming instructions, on-chip memory cells to provide data, and a logic unit for manipulating the data.

Another Japanese firm succeeded in producing a circuit with the fastest logic speed ever attained. Fujitsu scientists built and operated a simple series of Josephson logic subcircuits averaging a mere 2.5 picoseconds per switch. Remarkably, emerging materials and new circuit ideas threaten even this incredible record.

Progress on large-scale Josephson circuitry continues in earnest. Predictions of when a complete Josephson computer will appear have been notoriously inaccurate. But even the most conservative observer would agree that the performance of ultrafast Josephson logic has passed the test of medium-scale integration. MITI's Josephson computer may be only a few years away. Meanwhile, Josephson electronics has plenty to offer beyond high-speed computing.

From Analog to Digital

As anyone with a new stereo system can attest, digital electronics has been making inroads on the formerly exclusive domain of analog electronics. The inherent accuracy of digital data and the ease with which it can be handled has prompted the use of digital techniques in an increasing number of noncomputer applications. Thus we now have digital compact discs, digital television, digital radar, and digital telephone systems. But all these applications point out an ironic fact: the original sources of information—like light, sound, and radio waves—are all analog in nature; they don't start out as lists of numbers. So even if digital electronics takes over virtually every former stronghold of analog electronics, there will always have to be one device to bridge the gap between the two kinds of electronics: the analog-to-digital converter. And even this specialized device can benefit from superconductivity.

We have seen that digital circuits process information in the form of binary numbers—bits—rather than as time-varying signals. Moving bits around can be accomplished with far greater accuracy than can be achieved with complicated waveforms. But the numbers have to come from somewhere. Analog signals have to be converted into digital signals in order to work with digital circuits. Analog-to-digital converters—usually called *A/D converters*—perform this task in today's increasingly digital world. A/D converters can be bought off the shelf at any electronics store. In circuits like digital music recorders and video processors, current A/D technology handles the job with ease.

Where does superconductivity enter the picture? Just as in the case of computers, superconductive electronics will be called upon when the limits of semiconducting technology must be stretched and finally broken. A/D converters face two performance limits: speed and resolution. The speed of a converter corresponds to how often an analog signal is sampled, just like the frame rate of a motion picture. The resolution of a converter determines how accurately it can reproduce a signal; in conventional terms, resolution corresponds to the number of decimal places in a measurement. The challenge in A/D design comes from the fact that speed and resolution always compete; faster converters have intrinsically lower resolution and very accurate converters are inherently slower.

Customers for ultrahigh-performance electronics—like satellite designers and high frequency communications engineers—can no longer accept this speed-resolution tradeoff. The most advanced electronics systems on the drawing boards require A/D converters that are both extremely fast and extremely accurate. Engineers need converters that sample a thousand times faster than today's devices and they need resolution in thousandths of a percent. Such resolution corresponds to having a 12-inch ruler that reads in increments of ten-thousandths of an inch. When such performance requirements are presented to semiconductor engineers, they begin to get uncomfortable. When customers add that they want to place thousands of A/D converters on board a satellite with limited electrical power, semiconductor engineers head for the nearest exit. Superconductive electronics provides the only answer.

A/Ds need to convert ordinary electrical signals into binary digits. There are several superconductive electronics schemes to do the job, including one called *single flux quantum* (SFQ) logic. This superconducting technique exploits the behavior of a SQUID—a superconducting loop containing a weak link. As we recall from our earlier discussions, every time we try to put too much current into a SQUID loop, the weak link ceases to allow supercurrent to pass and a voltage briefly appears. By counting these bursts of voltage, the SFQ circuit can produce a digital representation of an analog signal fed into the SQUID. The SQUID circuit behaves like a turnstile at the entrance to a stadium. Just as a turnstile records a higher number each time a fan enters the stadium, an SFQ circuit produces and tallies voltage pulses for every increment of incoming signal. The details of A/D converter operation get very complicated and are really only of interest to circuit specialists. What matters is that SQUIDs can respond with tremendous speed to changing analog signals and, like all superconductive circuits, can do the job using a thousand times less power than equivalent semiconducting circuits. Superconducting A/D converters are not likely to show up in consumer electronics over the next few years, but they are destined to have a major impact upon advanced electronic systems of the future.

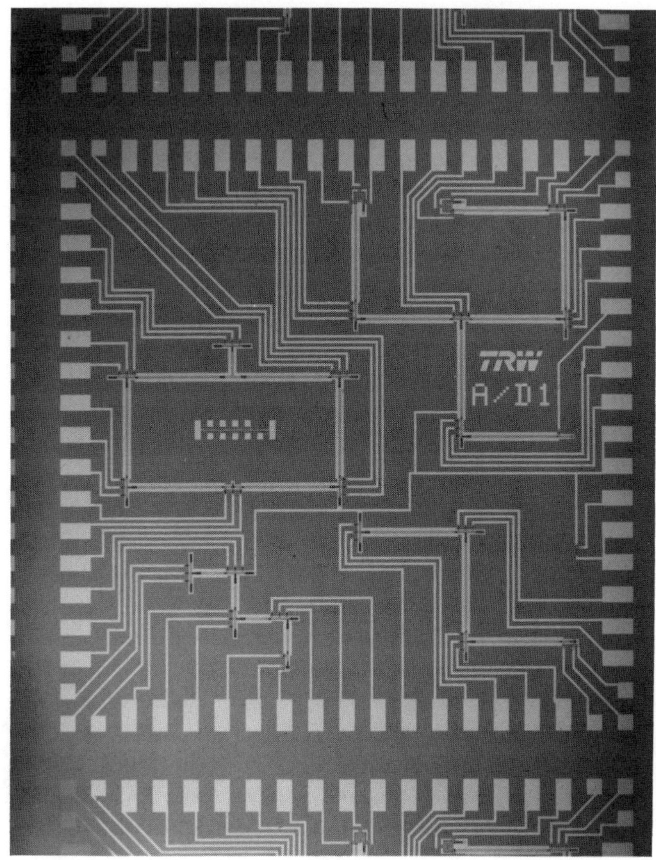

A Josephson electronic circuit. (Courtesy TRW)

Sampling the Nondigital World

Superconductive electronics has cut its teeth on digital applications like computer circuits and A/D converters, but the same techniques that produce these circuits can also be used for high-performance analog electronics. In fact, analog applications are proving to be superconductivity's initial entry point into the electronic marketplace. The high-speed, high-sensitivity characteristics that excited digital designers can also be harnessed in analog circuits. Several of the initial analog projects, in fact, evolved directly from digital circuits and designs.

Modern electronics continues to get ever faster. Even semiconductor circuits can now function on picosecond time scales. For the engineers who design and work with these circuits, a peculiar problem has surfaced. The circuits are so fast that no test instruments exist that can keep up with them. Engineers cannot assess the performance of their own devices. Once again, superconductivity comes to the rescue, this time in the form of an instrument called the *sampling recorder*, or *sampler*.

Samplers deal with ultrafast periodic signals by measuring them in a piecemeal fashion. Instead of painting a picture of the entire signal at one time, samplers piece together a time mosaic. Taking advantage of the recurring nature of the signal, the instrument "samples" one little time slice of the signal per measurement. If enough sensitivity and precision can be maintained in the individual measurements, the final composite picture accurately describes the high-speed input signal.

Josephson junctions provide a way to build high-speed sampler circuits. The Josephson sampler circuit uses a mousetrap approach to detect high-speed signals. As we know, a junction that carries the maximum amount of Josephson supercurrent will switch to resistive behavior when any additional current is applied. This super-fast switching forms the basis of computer circuit elements. In a sampler, a signal that lasts for even a few picoseconds can trigger the appearance of a voltage in a suitably prepared Josephson junction. So junctions are set like mousetraps to respond to fleeting signals. If the traps are set with the right timing, even the fastest signals can be "caught" as they fly by. The voltage that appears on the junction remains until the circuit is reset. By using a set of junctions, we can produce a snapshot of the passing signal that can be observed at our leisure.

During the heyday of the computer project, IBM researchers demonstrated a high-speed Josephson sampler and reported their results to the scientific community. A fair bit of scientific curiosity was aroused, and a number of other researchers came up with improvements. Finally one of the sampler's inventors left IBM and, armed with venture capital, formed a company dedicated to building and marketing Josephson electronics laboratory instruments.

The company, called Hypres (as in "hyper-performance research"), now markets a Josephson sampler. The complete sampler system takes up the space of a kitchen stove and includes the superconducting sampler chip (less than an inch square), a supply of liquid helium, a novel chip cooling apparatus, and all the conventional room temperature electronics required to tie the system together. One version of the Hypres sampler goes by the unwieldy name of the Time Domain Reflectometer, or TDR for short. The TDR helps circuit builders troubleshoot new high-speed circuits. When high-speed electronic signals travel through circuits, they bump into imperfections in wires and devices and broadcast minute radio frequency signals back through the circuit. These reflected signals can be used to electronically inspect circuits on a microscopic scale. The TDR that does the job acts like radar for wiring. A superconductive electronic pulse generator sends pulses of current into the circuit while the Josephson sampler sits and listens. Analyzing the signals that reflect back reveals imperfections in an electronic circuit in the same way that reflected radar signals pinpoint aircraft in the sky. The TDR places a powerful technique in the hands of the designers of high-speed electronics.

Amplifiers and Oscillators

We have seen how electronics has continued to evolve towards faster, more precise, and smaller circuitry. But it has also moved in the direction of higher frequency circuits as well. Engineers want to exploit ever higher frequencies for electronics applications, and superconductivity will be the way to do it. Semiconducting electronics faces fundamental frequency limits set by the properties of semiconductor materials. The frequency limits of superconductive electronics, on the other hand, are dramatically higher.

Superconductive electronics for high-frequency analog applications has proven to be one area in which complicated circuits containing hundreds or thousands of junctions are not always required in order to perform useful tasks. In fact, some of the highest performance analog circuits ever built made use of a single Josephson junction. Operating high-frequency analog electronics requires critical circuit building blocks like oscillators and amplifiers. Both of these functions can be accomplished with a single Josephson junction, or perhaps with a handful.

Circuits must produce their own high-frequency signals for

many purposes and oscillators perform that function. Earlier, when we discussed the ac Josephson effect, we saw that applying a voltage across a Josephson junction adds unwanted energy to the junction. The junction responds by casting off this energy in the form of electromagnetic radiation. A Josephson junction becomes a tiny radio station, broadcasting a signal whose frequency is precisely set by the voltage across the junction. In electronic engineering terms, a Josephson junction therefore behaves as a voltage-controlled oscillator (VCO). To produce a signal of a particular frequency in a Josephson VCO, we simply apply the correct amount of voltage across the junction. This makes for an elegant oscillator circuit.

Amplifiers are essential circuit elements because small electronic signals must often be made larger in order to be useful. Engineers rely on amplifiers to solve most circuit problems. If an incoming signal is too weak, then amplify it. If noise causes problems, amplify the signal until it surpasses the noise level. If losses weaken the output signal, amplify again. Experiments have demonstrated that Josephson junctions can be used as amplifiers. While the Josephson amplifier remains as a test device in the laboratory, its performance has been extremely impressive. For the electrical designer, the key advantages of the Josephson amplifier are its high-speed and low-noise operation. The circuits amplify well at very high frequencies and produce less noise than any semiconductor amplifier. In fact, Josephson "parametric" amplifiers continue to set performance records for high-frequency, low-noise amplification. Coupled together, Josephson amplifiers and oscillators can form the basis of advanced new high frequency electronic circuits. Both amplifiers and oscillators are currently under development for a number of high-performance applications in astronomy, communications, radar, and other areas.

The Josephson Volt

High-speed, high-frequency, and low-power operation do not motivate every application of superconductive electronics. Sometimes designers make use of the Josephson phenomena to tackle tasks that would be unapproachable with conventional electronics. Such a use is the Josephson voltage standard, an important application that has already placed Josephson devices in the field.

What is a voltage standard? A standard is a reference example for a unit of measurement. In the past, scientific congresses defined each unit by use of physical examples. A custodian kept a standard kilogram weight, for example, and allowed other people to balance their weights against it. As our needs for more precision grew, international standards committees decided that standards based on reproducible natural phenomena could be more precise than individual physical specimens and would be accessible to every laboratory. As an example, the standard meter is no longer defined by a bar of platinum locked up in a laboratory. Now the meter is chosen to be the distance that the characteristic light emitted from a krypton atom travels in 1,650,763.73 wavelengths. That may seem like a strange and complicated definition, but it provides a standard that can be set up anywhere, never changes, and turns out to be rather easy to measure in the modern laboratory.

Precision is the watchword of modern technology. Jet engines work only if turbine blades fit their housing precisely. Color television displays true colors only if frequencies lock exactly. Aircraft guidance works only if time is maintained accurately to the nanosecond. Measurements are so important that an entire branch of science, metrology, has developed to better our abilities to make precise and accurate measurements. In our electronic age, electronic standards for voltage and current rival the standards of length, mass, and time in importance. Now that there are instruments that can measure voltages to extraordinary precision, scientists need better ways to define the standard volt.

Until the early 1970s, batteries set the United States voltage standard. A very special bank of batteries was maintained and cared for at the U. S. National Bureau of Standards, in Gaithersburg, Maryland. As far as batteries go, these lived lives of luxury. To keep their temperature absolutely constant, they were immersed in a constantly stirred bath of high-purity insulating oil. A special electromagnetically shielded room protected them from the effects of radio waves, lightning, and other electrical noise. They even had a pedigree signed by the U.S. Congress. By legal act these cells defined the U. S. Standard Volt. All officially calibrated voltmeters in the United States maintained a calibration lineage back to these battery cells, the standard battery. Recalibrating a voltmeter required a painstaking procedure involving cross-comparisons with secondary and tertiary traveling standards—special batteries that shuttled back and forth between the Bureau of Standards and other laboratories.

But all these pampered batteries were not stable enough for today's ultraprecise electronic instrumentation. Engineers really needed a universal standard that could be easily maintained by any laboratory in the world. Batteries, with all their drifts and idiosyncrasies, just weren't cut out for the task.

The Josephson junction proves to be the standard that metrologists were looking for. The ac Josephson effect had provided scientists with a way to relate voltage to time. Modern electronic techniques had already provided excellent ways to standardize time and frequency measurements. While applying a voltage across a Josephson junction produces radio waves, exploiting this phenomenon in reverse produces a world-class voltage standard.

If we beam radio waves at a Josephson tunnel junction, a voltage will spontaneously appear across the junction. The junction replaces the standard battery, but the Josephson "battery" is better than any ordinary battery, since its voltage is as steady as a rock. The output voltage never wavers as long as the radio frequency stays the same. Furthermore, any two Josephson voltage standards will give the identical results every time. In short, Josephson junctions make an ideal voltage standard.

In 1972, at the urging of scientists at the National Bureau of Standards, Congress passed a new law. Henceforth superconductivity would provide the U. S. with its standard of voltage. So today, instead of sending out inexact secondary voltage standards to American laboratories, the Bureau of Standards can provide a Josephson package with all the accuracy of the original standard. A calibration standard consisting of a superconducting Josephson chip and a frequency standard can be set up anywhere in the world. Individual Josephson junctions produce only tiny voltages under the influence of radio waves of convenient frequency. To build a voltage standard of one volt, for example, several thousand Josephson junctions are required. Fortunately, by using the sophisticated circuit techniques that were developed for Josephson digital electronics, scientists can readily produce voltage standard chips containing thousands of high-quality Josephson junctions.

As we have begun to see, electronics has grown to be a field of almost bewildering diversity and complexity. In this chapter, we have tried to give a flavor of just some of the areas in which superconductive electronics already plays a role and where it may enter the picture

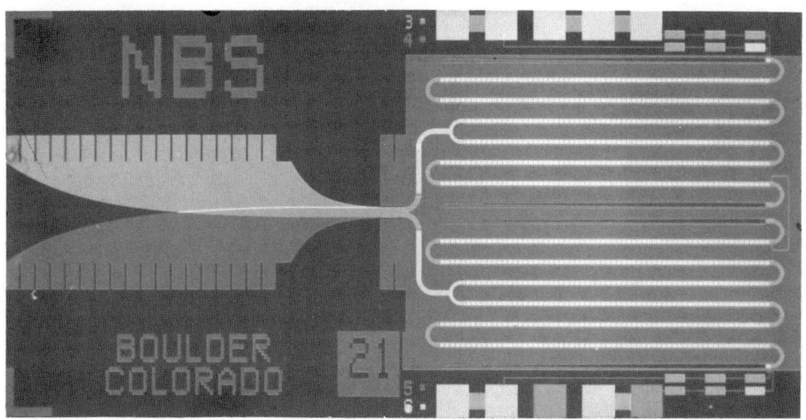

A Josephson voltage standard chip. (Courtesy National Bureau of Standards, Boulder)

in the future. A more comprehensive survey of electronic application could fill an entire book—and in fact, several such books already exist. For our purposes, we should come away with a clear picture that superconductivity has a great deal to offer to high-performance electronics. In the next chapter, we concentrate on yet another aspect of Josephson electronics. Again, it is a place where a single Josephson junction is all we need. We explore the world of SQUID magnetometry.

The extent to which Josephson integrated circuit technology will penetrate the electronic marketplace will depend strongly on advances in new materials, processing, and circuit developments. Nevertheless, Josephson circuits have undeniable selling points for a wide variety of high-performance circuits. Superconducting circuits are inherently fast, quiet, and precise, and they are here to stay.

CHAPTER 10

The World of SQUID Magnetometry

Superconductors have earned their fame with their electrical properties, but so far they have earned their living with their magnetic properties. The electrical properties of superconducting wire will undoubtedly play a growing role in a number of future applications, but superconductivity has already become indispensable for advanced uses of magnetism. Superconductivity allows us to generate immense magnetic fields for a variety of applications in science, medicine, and industry. And at the opposite extreme, superconductivity also allows us to detect the presence of incredibly minute magnetic fields in the earth, under the sea, and within the human body. Sensing magnetic fields requires the techniques of magnetometry, and superconductivity provides the best magnetometers in the world.

We perceive the world through the five senses: sight, hearing, touch, smell, and taste. How different the world would seem if we had an additional sense, one that detects magnetic fields. With this sense we would never lose our way; our natural compass would point the way to the North Pole. But a magnetic sense would mean much more than that. Seeing through mountains, instantly diagnosing diseases, and reading geological history in the rocks would be just a few of the abilities we would gain.

The Magnetic World

Before going into detail about how we would apply this new sense, or more practically, how magnetometers really do let us see into the magnetic world, we will first try to imagine what the world must look like magnetically. With our new magnetic sense, our first view of the world would look pretty drab and uniform. The earth's magnetic field steadily bathes everything around us in a uniform shower of magnetic field lines. Such field lines easily penetrate air, trees, concrete, and even miles of rock.

Only when we look a little more carefully do we start to notice differences in the magnetic fields surrounding different objects. Often the effects are subtle, but clearly some materials twist and distort magnetic field lines. Such effects not only save us from a rather dull landscape, but are crucial to many of the most important applications of magnetic sensors.

Different materials have different magnetic properties. If we examine the materials around us, we can categorize them according to what happens when magnetic field lines try to pass through them. If we were interested in their optical properties, we would sort materials according to what happens when light tries to pass through them. Materials would then be labeled "transparent," "opaque," and "reflective." For magnetic properties, we need a new set of labels. Borrowing some scientific terms, we would label materials "diamagnetic," "paramagnetic," and "ferromagnetic."

Many materials turn out to be diamagnetic. These materials expel magnetic fields. Recall that superconductors are perfect diamagnets; no magnetic field penetrates the interior of a superconductor. Such absolute diamagnetism sets superconductors apart from all other materials. Most diamagnetic materials make only a feeble effort to fight off fields and exclude only a tiny fraction of any applied magnetic field. Laboratory researchers measuring the diamagnetism of water, for example, find that it rejects only about one millionth of an applied field. The diamagnetic category encompasses the greatest number of natural substances, as it includes most plants and animals, as well as many elements and molecules.

A smaller class of substances falls under the heading of paramagnetic materials. These materials are magnetic sponges, actually attracting magnetic fields. The interior of a paramagnetic substance will concentrate lines of magnetic field from its surroundings. As with diamagnetism, paramagnetism is generally a weak effect, often amounting to magnetic changes of only a few parts in a million. Quite a few substances are weakly paramagnetic, including oxygen gas, most metals, and a number of iron-based compounds.

The final category contains fewer materials, but the behavior of the ferromagnets sets them well apart from the diamagnets and paramagnets. If diamagnetics expel magnetism, and paramagnets attract it, what is left for our third magnetic category? Ferromagnets not only strongly attract magnetic flux, but they can become permanently magnetized. Even after we remove the source of magnetic field, a ferromagnet retains its magnetization. Iron, nickel, and cobalt are the most familiar ferromagnets, and we use their ferromagnetic properties in building compass needles and horseshoe magnets. More powerful permanent magnets can be made from ferromagnetic alloys like alnico (containing aluminum, nickel, and cobalt) and recently developed rare earth alloys like samarium-cobalt and neodymium-ferrite.

Nearly all materials fall into one of these three magnetic categories. The three categories describe passive magnetic behavior: how a material behaves in the earth's field or any other external magnetic field. Some of the most useful applications of magnetic sensing depend precisely on these characteristics. But looking a little further, magnetic sensing can tell us even more about the world around us.

Even a cursory magnetic inspection of our environment reveals a more active form of magnetism. We can easily find many magnetic fields that have nothing to do with the earth's magnetic field. For example, electrical appliances all produce their own detectable magnetic fields. Any electrical equipment in which currents flow must generate magnetic fields, but these are not the only currents that produce magnetic fields. Even biological currents from nerves and muscles are magnetic sources. Our bodies sparkle with magnetic fields radiating outward every time a nerve ending fires or a muscle contracts. Any place an electrical current flows, a telltale magnetic field will also appear. Detecting and interpreting the fields from these active sources leads to a second series of magnetic sensing applications.

So far the magnetic inspection of our surroundings has turned up *passive magnetization*—appearing as diamagnetism, paramagnetism, and ferromagnetism in the earth's field—and *active magnetization*, caused by flowing currents. A third source of magnetic fields, *reactive magnetization*, plays an equally important role in applications of magnetic sensing.

We have been talking about steady magnetic fields. Reactive magnetization, on the other hand, deals with the effects of changing fields. Lenz's Law of induction tells us that changing magnetic fields produce electrical currents in conductors. Good conductors and rapidly changing fields produce large currents. However, changing currents and changing fields have a chicken-and-egg relationship. If a changing magnetic field causes currents, and currents in turn cause new magnetic fields, then a changing magnetic field must ultimately generate new magnetic fields of its own. It's a simple matter of action and reaction. We call this process reactive magnetization, and physicists find it quite useful in measuring the electrical conductivity of a material without having to touch it or even come anywhere near it. Geologists even have a scheme for finding oil using reactive magnetization.

With all these types of magnetization around us, the world could reveal a richness of detail if only we had a magnetic sense. Although we are physically limited to the five senses, a little electronic ingenuity and some help from superconductivity can make up for nature's oversight. Magnetometers, special sensors that accurately measure magnetic fields, are our eyes for looking at the magnetic world. Many different magnetometer designs crowd the electronic marketplace, but for the greatest sensitivity, scientists, engineers, and doctors turn to superconducting magnetometers.

SQUIDs

Whereas the techniques for constructing magnetometers and extracting information from them have become sophisticated and complex, the basic concepts remain quite simple. Superconducting magnetometers take advantage of flux quantization with the use of SQUIDs (Superconducting Quantum Interference Devices). Remember that a superconducting ring can only enclose magnetic fields from a specific set of values. Specifically, the magnetic flux in the ring—the product of the magnetic field and the area of the ring—can only take on values that are multiples of a specific number: the flux quantum. Thus a ring may have one, two, or a thousand flux quanta, but it can never have one and a half. As a result, when we place a superconducting ring in a magnetic field, a current flows around the ring to keep the flux in the ring at a permitted value.

SQUIDs were named and first studied at the Ford Motor Company Scientific Laboratory in Dearborn, Michigan, in the mid-1960s. There a group headed by Arnold Silver and James Zimmerman performed the first experiments on the magnetic response of superconducting loops, discovering that these structures displayed extraordinary sensitivity to small magnetic signals. The earliest SQUIDs were handmade devices fashioned from tiny loops of superconducting wire. The Ford group was able to develop a model for the behavior of their devices that forms the basis of our current understanding of the SQUID.

SQUIDs are fundamentally superconducting rings that act as storage devices for magnetic flux. In building SQUIDs, engineers intentionally incorporate weak links in the superconducting rings. These days the weak links are most often Josephson tunnel junctions, which can now be made reliably and reproducibly. In a SQUID, the weak link provides us with a way to change the amount of flux in the ring. It acts as a flux gate. If we apply an outside magnetic field, flux will want to enter the SQUID ring. If we decrease the magnetic field, flux will attempt to leave. Ordinarily, the superconducting ring will oppose any change of flux by generating superconducting currents. Adding the weak link introduces another option. If we apply too large a magnetic field, the current induced in the ring will exceed the maximum current that the Josephson junction can handle. At that point, the superconducting weak link will temporarily fail (a voltage appears across the junction, breaking the continuity of the superconducting ring), and additional magnetic flux can enter the ring. If the excess field was just a little too large, then the junction will regain its Josephson current, the ring will be complete again, and there will be one more flux quantum threading it. If the field was somewhat larger, then perhaps two or more additional flux quanta will be stored when the junction regains its superconductivity.

We can turn this simple SQUID ring into a device for measuring magnetic fields. Engineers can electronically detect how much current flows through the weak link at a given time. A technique called "feedback" works by intentionally adding just enough extra current to the ring to cancel out any currents generated by magnetic fields. This keeps the flux constant in the SQUID. Knowing the current we add to the ring to balance out the induced current tells us the amount of magnetic field sensed by the SQUID. The technique is similar to weighing an object with a pan balance; putting known weights on one pan until both pans balance tells us the weight on the other pan. A good balance can detect tiny differences in weight between two objects. Similarly, clever electronic feedback circuits for SQUIDs enable engineers to detect magnetic field strengths that correspond to less than a thousandth of a flux quantum in the SQUID ring.

Sensitivity to tiny magnetic fields is the key to the preeminence of superconductivity in magnetometer applications. SQUIDs can detect small fractions of a flux quantum, and a flux quantum can correspond to very small magnetic fields. Consider the example of the earth's magnetic field. Detecting this field by mechanical means is not a trivial task. One needs a well-balanced compass needle to track the earth's North Pole. For a SQUID, on the other hand, the magnetic field of the earth is enormous. A SQUID loop the size of your little finger would enclose about a million flux quanta when threaded by the earth's magnetic field. Today's commercial SQUIDs can easily measure magnetic fields billions of times smaller than the earth's field. This unmatched sensitivity overcomes our lack of a biological magnetic sense. It also opens the door to a wealth of applications ranging from prospecting to submarine hunting to medical diagnosis.

Scientists and engineers from many different disciplines are devising a number of applications for superconducting magnetometry. Laboratory scientists have used commercially produced SQUID magnetometers since the early 1970s. The passive magnetic behavior of a variety of materials has been studied with the help of SQUIDs. Today SQUID magnetometers are playing an important role in the further development of superconducting materials by measuring the diamagnetism—a key superconducting property—of the new ceramic superconductors. But the greatest interest in superconducting magnetometry lies in applications outside of the laboratory. Here SQUIDs are finding a broad variety of uses ranging from sensing geological changes deep within the earth's crust to probing the activities of the brain.

Sensing beneath the Earth and Sea

One of the most straightforward applications of SQUID magnetometry may turn out to be the search for valuable ore deposits. Prospectors may someday rely on SQUID magnetometer maps to guide their searches. Typically, miners know the kind of geological layers in which minerals are located but often these important strata are hidden from view by layers of surface rock.

Magnetometry can overcome this problem because different rock layers show different diamagnetic or paramagnetic responses to the earth's magnetic field. Igneous and metamorphic rock—originating deep within the earth—tend to have moderately high concentrations of iron-rich magnetite, as opposed to the more diamagnetic sedimentary rock. As a result, transitions between different strata produce detectable changes in the earth's magnetic field that we can measure. Airborne experiments in Canada have successfully mapped geological magnetic features using nonsuperconducting magnetometers. Superconducting magnetic sensors promise to give map makers an extra factor of 10 in sensitivity, allowing more detailed mapping to be conducted in less time.

These magnetic maps, and especially changes in the maps over time, may prove useful beyond just prospecting. Earthquake scientists study magnetic changes caused by stresses and shearing motion in rock formations. Scientists hope someday to predict likely earthquake periods by magnetically monitoring strain buildup deep beneath the earth.

Back at the lab, geologists also find uses for SQUID magnetometers. Many of these scientists rely on superconducting magnetometers to analyze the rock samples they uncover. Of course, chemical assays tell them what atoms or chemicals are present in a sample, and sometimes that is all the information they need. Magnetic sensors, in contrast, tell scientists how the atoms in a substance combine to form magnetic structures. Special geological magnetometers not only measure the magnetization of rock samples, but show how the magnetization changes with temperature. With this detailed knowledge, scientists can infer the composition and amount of magnetically active materials. Unlike many kinds of chemical testing, the magnetic analysis is nondestructive. That might not mean a whole lot for common ore samples, but with something like moon rocks, scientists have a better chance of borrowing the sample if they can return it in one piece.

Of all the fields of prospecting, none offers greater rewards nor employs more people than the hunt for oil. Oil geologists are willing to employ very active search techniques. Exploration for oil will continue to be a required chore to support our mechanized society, and SQUID magnetometers may help us in the search.

SQUIDs are well suited to paying their way in the oil business. Finding oil is always risky. Oil companies have to invest money for a drilling rig and a crew to sink a well deep into the earth, and they only have a finite chance of striking an oil reservoir. If the well comes up dry, they have nothing to show for their money and effort. Obviously, life would be much easier if we could see through the layers of bedrock and shale and know whether the oil is there in the first place.

Toward that end, engineers apply the most basic ideas of reactive magnetism to probe deep below the earth's surface. To be specific, geomagnetic sensing measures the conductivity of rock layers deep within the earth. All that's required is a very sensitive magnetic detector and a changing magnetic field.

Searchers actually have two sources of changing magnetic fields at their disposal. Geologists haul large magnetic coils to a site and energize them with generators in order to produce strong magnetic fields at the earth's surface. The field strength dies off away from the magnets, but such intense magnetic fields can reach down the several kilometers needed for geological exploration. Alternatively, the geologists can use natural magnetic fields to probe deep within the earth. As every scout with a compass knows the earth has its own magnetic field, built in. We think of this magnetic field as being pretty steady, and it is, but not perfectly so. As the earth spins on its axis, the solar wind (actually a stream of charged particles from the sun) exerts a kind of magnetic pressure on the earth's own field, pushing it around a little. Electrical current from lightning strikes also produces much faster changing magnetic fields that penetrate deep underground. One way or another, nature provides changing magnetic fields that probe every inch of ground over the entire planet. To take advantage of these probes, all we have to do is build a detector sensitive enough to read the results.

By setting up SQUID magnetometer stations at ground level, scientists can correlate the detected magnetic signals with the applied magnetic field. Low-frequency magnetic fields—fields that change slowly—penetrate more deeply into the earth than high frequency fields. By computer analyzing signals at many different frequencies, geologists map out the conductivity (and hence composition) of deep underground layers. The whole operation yields detailed geological information without having to move a shovelful of dirt.

Economics decides the commercial viability of magnetic oil exploration. Magnetometry complements other, more common geological explorative techniques such as acoustical sounding. For now, with oil prices low, the latter technique carries the market. As oil prices rise, economic forces will push for better locaters of harder-to-find oil.

Expect superconducting magnetometers to emerge into more wide-spread use in the near future.

The magnetic response of materials has created yet another application for superconductivity in the ore refining industry. Refiners can use powerful magnets to pull paramagnetic impurities out of ore. This magnetic separation technique has proven to be very valuable in the processing of fine clay for the manufacture of pottery, paint, and glaze coatings. Commercial separators are now available that make use of superconducting magnets.

Similar magnetic separation techniques can be applied to the problem of water purification. Engineers intentionally add iron oxide particles to unpurified water. Suspended solids, bacteria, and other contaminants attach themselves to these magnetic hosts. The superconducting magnets in a separator can then pull off a large fraction of the unwanted contaminants, leaving behind purified water.

Geologists prospecting for likely ore or oil sites use magnetometry to "see" through otherwise impenetrable solid rock. Imagine what happens when hunters of a different sort stalk a prey equally well hidden, but mobile as well. That's exactly the problem faced by naval commanders on all the oceans of the globe as they try to locate unfriendly submarines.

Finding a submerged submarine is no easy job. Radar only detects objects above the surface. Visual searching doesn't do much better. Neither radio waves nor light penetrates more than a few meters into seawater. Acoustical techniques like sonar and passive listening work a little better, but aren't ideal. Sound travels further than light or radio waves underwater, but changes in water density reflect back sound waves, limiting the effectiveness of acoustical searches. The problem boils down to one of transparency. When it comes to light, radio waves, and sound, seawater just isn't transparent enough for sub hunters.

Magnetic detection offers searchers an advantageous alternative to sonar. To magnetism, seawater remains crystal clear. Today's submarines are big and have lots of iron-laden steel in them. As a result, the presence of tons of ferromagnetic material strongly distorts the earth's magnetic field as a submarine passes by. To a magnetic eye, a submarine shines like a beacon.

Airborne, ship-towed, and sea-bottom magnetic sensors are all conceivable, although each has technical difficulties that engineers will have to face. To date, not much has appeared in published literature about the role of magnetometry in undersea warfare, but certainly the advantages of such uncloakable detection have caught the eye of Pentagon planners and their overseas counterparts.

Ironically, superconducting magnetometers also prove useful to submarines, although in an entirely different way. One of the most dangerous times for submarines is when they come close to the surface to receive radioed orders. The same seawater bed that cloaks the submarines also blocks ordinary frequency radio messages. In order to allow submarines to remain deeply submerged while receiving orders, the Navy broadcasts very low frequency electromagnetic waves. Submarines can magnetically detect these waves using SQUIDs. The transmission procedure resembles the techniques used in oil prospecting, but in this case the receiver is buried underwater.

Not all ocean-going SQUID magnetometers are in the hands of the military. Scientists who trace the movements of the earth's magnetic poles take to the seas as well. In this case, the scientists hunt a historical prey that is millions of years old.

The oceans allow the earth to constantly lay down a magnetic record of its history. A gentle rain of particles constantly settles from the seawater onto the ocean floor. A few of these particles are ferromagnetic and thus have their own magnetization. As they fall, these magnetic particles act like tiny compasses. More often than not, when they hit the ocean floor they align with the earth's magnetic field. Once covered with more material, these tiny particles become locked in place.

Superconducting magnetometers provide the sensitivity required to read back the magnetic history so conveniently laid down for us. Ocean-going geologists retrieve core samples of material laid down over the course of many millions of years. SQUID magnetometers not only tell the strength of the trapped magnetization in these samples, but the exact directions as well. The shifting orientation of these tiny compasses pinpoints the changing direction of the earth's magnetic poles over time. As a result, scientists can reliably track the movement of the earth's magnetic poles over geological time scales. Archaeological samples can provide similar information for more recent times. Certain iron-rich clays, often found in kiln material, are paramagnetic at oven temperatures, but lock in a remnant of the earth's magnetic field as they cool. Once again, SQUID magnetometers can decipher these trapped magnetic messages to reveal the earth's magnetic history.

Magnetism and Medicine

Intriguing as all these applications may be, none promises greater gifts to humanity than the biological uses of magnetometry. When fighting diseases, doctors have always faced the challenge of diagnosing what's wrong before effective treatment can begin. Superconductive magnetometers are stepping in to create new diagnostic procedures that are faster, more accurate, and less painful.

The basic idea of magnetically probing for disease has a certain attractive logic to it. The action of human maladies tends to hide within the afflicted organs. With the exception of skin diseases, doctors have no convenient way of really seeing what is wrong without opening the body. Whereas optical inspections must end at the patient's skin, magnetic probing effortlessly penetrates the deepest places that diseases strike. Perhaps no population would benefit more from magnetic sensing technology than the medical profession and their patients.

One of the earliest medical applications of superconducting magnetometers relies on a simple application of what we know about passive magnetization. The target disease is *hemochromatosis*, a genetic disorder afflicting roughly a million Americans to one degree or another. Despite its prevalence, the disease often escapes detection. It is difficult to diagnose, often overlooked, and potentially life threatening.

Hemochromatosis involves an imbalance in the body of one vital element: iron. Iron is essential for building blood cells and carrying oxygen throughout the body. A healthy body regulates the amount of iron absorbed from foods and stores some of it in the liver. Our bodies don't need very large amounts of iron. The 50 to 500 parts per million found in the liver suffice to replace the hemoglobin and blood cells that wear out in our bodies every day. Any extra unneeded iron in our diets ordinarily flushes away with other bodily wastes.

Sufferers of hemochromatosis have lost the ability to regulate iron in their bodies. Levels of iron 20 times normal can accumulate in the liver of a hemochromatosis victim. Eventually the liver becomes saturated, releasing large amounts of iron into the bloodstream. In the blood, excess iron eventually interferes with the action of the heart, causing an unexplained heart attack in the undiagnosed hemochromatosis victim.

The control of hemochromatosis becomes relatively simple once accurate diagnosis has been made. Doctors control mild cases by placing the patient on a restricted diet; in other cases, hemochromatosis sufferers rid themselves of excess iron by donating blood with great frequency. Neither course of treatment is very complicated; the tough part of the job is making the correct diagnosis. Testing for iron ions in the blood is unreliable and often only shows signs when the disease has reached its final, most dangerous stage. Liver biopsies, involving removing and chemically analyzing a plug out of the liver, are more accurate, but subject the patient to an extremely painful procedure. Unless physicians strongly suspect hemochromatosis, from the patient's genetic history, for example, they hesitate to subject a patient to such a possibly needless trauma.

Superconducting magnetometers give doctors the sensitive, nonsurgical diagnostic equipment they need. The iron held in the liver is weakly paramagnetic. While the level of magnetism is far too small to detect by a crude method like holding a permanent magnet in front of the patients' torso, doctors can detect iron concentrations quickly and easily with an instrument called a *superconducting susceptometer*. A susceptometer combines a superconducting magnet with a superconducting SQUID magnetometer. The superconducting magnet applies moderately large magnetic fields to the area of the patient's torso. SQUID magnetometers then readily detect the paramagnetic response in the liver, giving doctors an accurate readout of iron in the body. The all-superconductor combination of large field magnet and sensitive detector makes for an ideal detection scheme for excess iron concentrations.

SQUID susceptometers are an example of how magnetic fields can be fashioned into convenient medical tools. Both the applied fields from magnets and the paramagnetic or diamagnetic response fields penetrate the body easily and safely. These qualities contrast with alternative diagnostic techniques that doctors have relied upon for years. Unlike high-energy X rays, magnetic fields do not damage cells. Unlike biopsies, magnetization testing does not require penetrating the skin and opening the body to risks of infection, an especially dangerous factor in hospital environments.

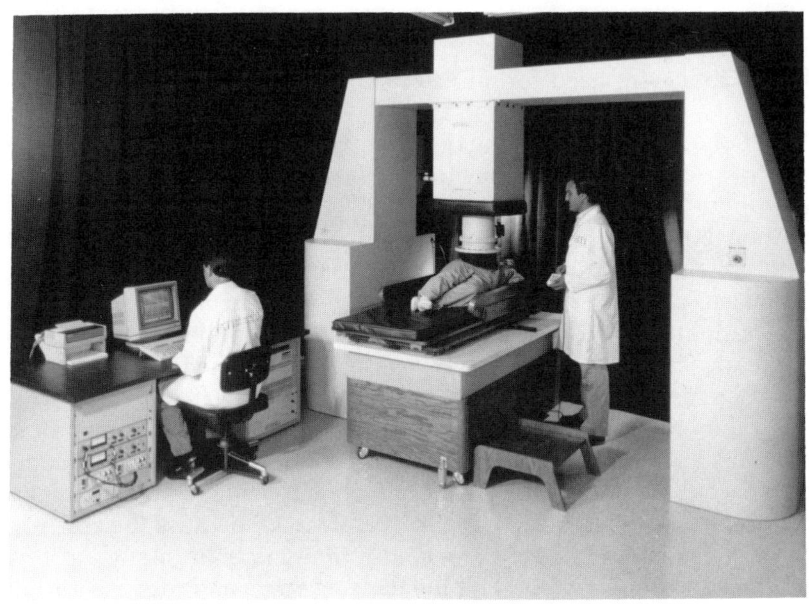

Magnetic sensing of liver iron concentration, using a superconducting magnet and SQUID sensors. (Courtesy Biomagnetic Technologies, Inc.)

The magnetic test for hemochromatosis is simple, accurate, painless, and noninvasive. Someday soon maybe all standard physicals will include a trip in front of a superconducting susceptometer.

Researchers use techniques similar to those of hemochromatosis diagnosis to study ferromagnetic iron contaminants in the body. Welders expose themselves to microscopic airborne particles when they fuse metals. The tiny particles of steel stick to lung surfaces. While such iron contamination of the lungs doesn't represent a major hazard to an appreciable segment of the population, tracking the movement of contaminants in general does interest physiologists. Using superconducting detectors, researchers can follow the cleaning action of the lung as it rids itself of the magnetically detectable particles. This gives the researchers a good picture of the body's response to nonmagnetic particles such as tobacco smoke or air pollutants as they are cleansed from the body.

So far we've seen the medical applications of SQUID magnetometers and susceptometers resulting from passive magnetization.

A large superconducting magnet sets up a field, and we read the response. For some diseases, an active magnetic signal already awaits us, without the need for an external field. Epilepsy is just such an ailment.

Epilepsy is a debilitating disorder. Sufferers of epilepsy face a life of drug dependency and inevitable side effects from their medication. Recently researchers have isolated the cause for epileptic seizures. Either through a head injury or a congenital condition, a portion of the epileptic's brain malfunctions. During an epileptic attack, the damaged portion of the brain essentially short-circuits. In mild forms epileptic attacks produce a loss of concentration. More violent attacks generate seizures as nerve pathways become jammed with meaningless signals from the damaged site.

The only known permanent cure for epilepsy is to remove the damaged portion of the brain responsible for the disorder. Here doctors face difficult decisions. All brain surgery endangers the patient. No doctor wants to risk having to operate twice because the epileptic center was missed in the first attempt. Enough brain tissue must be removed to completely excise the damaged area and prevent further seizures. On the other hand, extraction of too much brain tissue can create even more serious problems. Mental and physical impairment result from cutting key brain pathways. Neurosurgeons need to pinpoint the epileptic center precisely if they are to successfully effect a cure.

Superconducting sensors provide doctors with a much-needed tool to localize epileptic centers deep within the brain. The short-circuiting action of an epileptic center produces electrical currents that generate a distinctive magnetic signature. Doctors place an array of a dozen or so SQUID magnetometers around the patient's head. Computers analyzing the data from the different sensors can piece together a three-dimensional picture of activity within the brain. The technique, known as *magnetoencephalography* (MEG), has started to appear in leading medical centers. Commercial instruments built by a small company named Biomagnetic Technologies provide medical researchers a complete MEG system for something like $1 million.

MEG sensors outperform the electrical sensor grids that doctors previously relied upon. While an epileptic center generates both electric and magnetic fields, doctors can more readily locate it precisely with magnetic techniques. The skull represents a major difficulty for electric-voltage measurements on the brain. The solid bone masks

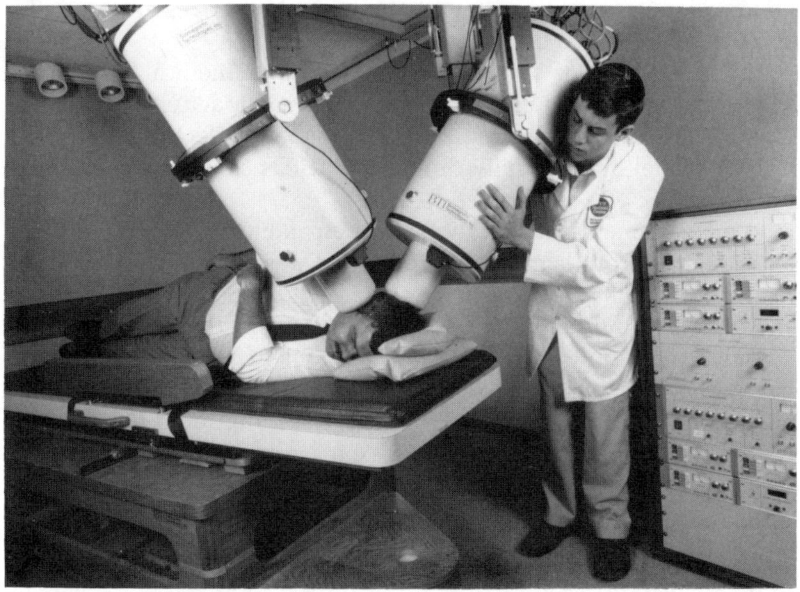

A magnetoencephallography system for precisely locating regions of brain activity. (Courtesy Biomagnetic Technologies, Inc.)

and smears the electrical signals from inner regions of the brain. Doctors must actually drill a large number of holes in the patient's skull to implant electrical probes within detection range. Worse yet, several sets of measurements are required in order to localize the epileptic center. Such surgery requires extended patient hospital care and introduces additional risks of infection.

Magnetic probing, on the other hand, turns out to be conceptually much simpler. Magnetic fields pass through the skull as if it weren't there. The three-dimensional nature of magnetic fields provides more information than the simpler voltage measurements. Doctors can nonintrusively search for epileptic centers using an array of SQUID sensors surrounding the patient's skull. Using such techniques, researchers can localize the brain's epileptic centers to within a few millimeters.

MEG also maps healthy brain tissue and describes its functions. By stimulating different parts of the body, scientists can trace magnetic signals backwards to match tissue location to function. The technique will never enable us to read thoughts, but it does give neurologists a new look into brain activity.

MEG techniques prove effective for other medical probing as well. German researchers have recorded magnetic activity in brains of fetuses. The mother's surrounding tissue makes electrical and other probes of fetal brain activity difficult. SQUIDs can sense even these tiny signals.

One final medical application of superconductivity has, within this decade, gone from laboratory curiosity to widespread hospital diagnostic tool. *Magnetic resonance imaging* (MRI) has reached the stage of both medical and commercial credibility. An increasing number of major hospitals now have MRI machines to help doctors with medical diagnoses. MRI acts somewhat like an X-ray machine, able to paint pictures of the human body, but with some important advantages.

The principles of MRI differ a little from the other magnetometry techniques so useful in iron susceptometry and magnetoencephalography. In fact MRI isn't really a form of magnetometry at all, although MRI does depend on superconductivity. Instead, MRI relies on magnetism and a special kind of behavior of atoms in a magnetic field.

The nuclei of many atoms, notably hydrogen atoms, act like tiny magnets. When an outside magnetic field appears, these magnetlike nuclei tend to line up with the field like a compass needle. Some atomic nuclei try to point the other way, but orienting against the magnetic field is not energetically favorable.

The signal used for MRI imaging comes about from the dynamic action of these nuclear magnets. When an atomic nucleus switches from being aligned *against* the magnetic field to being aligned *with* the magnetic field, a small amount of energy is released. The energy takes the form of an electromagnetic (radio) wave. The frequency emitted depends on the strength of the magnetic field and the particular kind of atom that emits the energy.

The MRI imaging process begins by placing the patient in the bore of a cylindrical magnet. The magnet maintains a steady, precise magnetic field through a cross-section of the patient's body. The magnetic field aligns the nuclei of the body's atoms and sets the frequency of the electromagnetic energy they emit. Engineers then supply a sharp electromagnetic pulse—a brief burst of energy—to kick these atoms. This temporarily scrambles many of the atomic alignments with the magnetic field. After the pulse has ended, the nuclei can realign with the magnetic field, and when they do they emit their characteristic radio signal. Arrays of detectors pick up the emitted radio waves for computer analysis. It takes technicians about half an hour to collect data for a good MRI scan. Sophisticated computer analysis then translates the data into a complete picture of a magnetic slice through the body.

An MRI scan corresponds to taking an electronic census of the body's hydrogen atoms. Upon command, every hydrogen atom lines up and broadcasts its location to tuned receivers. It takes a little cleverness to unravel all the data, but in the end MRI results resemble census figures. Diagnosticians end up with a map showing areas of high hydrogen population in the body.

Hydrogen turns out to be a very useful monitor of the body's health. Since two-thirds of water consists of hydrogen atoms, water-rich tissues respond strongly to MRI probes. Most organic materials are also hydrogen rich and also show up readily in MRI scans. In contrast, bones remain relatively dark. That's a big plus for doctors trying to examine brain tissue otherwise hidden by the skull.

Making magnetic resonance imaging practical for hospital environments takes careful engineering work. MRI system designers must satisfy tough demands and face some of their greatest challenges in the magnet system. For MRI to work, the magnetic fields must be strong, very uniform, cover an area large enough to span the patient, and maintain stability to better than one part in a million even over periods as long as an hour. Superconducting magnets, with their persistent currents' ability to hold fields absolutely constant, answer the demands made of MRI equipment. Because superconducting magnets are more compact and lightweight than their normal metal counterparts, superconducting MRI systems can fit into available hospital space without requiring big investments in special facilities.

In operation, MRI complements X rays, but offers some important advantages. Because bones block X rays, doctors still prefer X-ray images to study bone fractures and skeletal deformities. MRI, on the other hand, resolves organs that are nearly transparent and featureless to X rays. Tissue abnormalities show up strikingly on MRI images. Furthermore, doctors using MRI don't have to cope with the cell-damaging ionizing radiation of X rays. All in all, in a very short period,

MRI has emerged as a key medical diagnostic tool and superconductivity has played an important enabling role in its development.

As we have seen, despite the lack of an innate magnetic sense, scientists have conceived of a wide variety of ways to use the behavior of magnetic fields. Magnetometers using superconducting quantum interference devices have given engineers the tool they need to turn these ideas into reality. Magnetic resonance imaging, magnetoencephalography, and iron susceptometry provide doctors with valuable diagnostic tools in their war on disease. Ubiquitous magnetic field lines probe the deepest oceans and under the highest mountains for us. Medicine, geology, and electronic warfare will all make increasing use of the sensitivity that superconductivity alone can provide in detecting and analyzing magnetic fields.

CHAPTER 11

Superconductors in Scientific Research

Since the 1950s, scientists have struggled to develop superconducting materials and devices for an impressively diverse array of applications. Some of these endeavors have already yielded practical results, while others still face years of further research. But as experts continue to work to develop commercial, industrial, and military applications of superconductivity, their colleagues from other scientific fields already reap some of the fruits of their labors. The effort spent to develop superconductivity has paid off with large dividends for purely scientific research. Particle physics, astronomy, microwave engineering, and many other fields have all benefited tremendously from the advantages of superconducting technology. Superconducting magnets, materials, and devices have all become a part of modern scientific experiments. In this chapter, we explore some of the work being performed at the frontiers of science—work whose progress depends upon novel applications of superconductivity.

The Subatomic World

Over the years, no branch of science has grown to be more dependent upon superconductivity than experimental high-energy physics—the science of the workings of the subatomic world. High energy physicists seek to understand the nature of the various particles that make up the atom, as well as the mysterious forces that hold the particles together and govern their behavior. Ironically, physicists try to gain such understanding by working with powerful machinery built on a scale that dwarfs almost anything else of human creation.

The secrets contained within the atom are locked very tightly inside. Only violently destructive forces can loosen the array of subatomic particles. To accomplish this, physicists make use of giant particle accelerators, popularly known as atom smashers, and the latest atom smashers are big-time users of superconductivity.

The basic premise of a particle accelerator is quite simple. At the center of the atom sits the nucleus, a tightly packed bundle of protons, neutrons, and other particles. To find out what is inside the nucleus, we need to blow it up. This sounds like a brute force approach to a sophisticated problem, and it is, but atom smashing constitutes the only way we have to explore the inner workings of matter. Physicists also use particle accelerators to learn about subatomic particles by smashing them into each other. What comes out of such collisions can be surprising.

In the simplest picture, a particle accelerator works like the opening shot in a billiard game when the cue ball scatters all the other balls in a violent collision. In the accelerator, a particle like a proton is fired at enormous speed into an atomic nucleus, scattering its constituents. However, the billiard analogy breaks down with the addition of Einstein's famous law that equates energy with mass: $E=mc^2$. According to Einstein's theory, if a particle carrying enormous amounts of energy collides with another particle, new matter can be created. In other words, what comes flying out of the collision might not just be the participants in it; the collision can create entirely new particles. By studying such new particles, physicists gain important insights into the fundamental properties of matter.

There appear to be an almost inexhaustible supply of elementary particles. Physicists continue to develop theories that account for the ones that have been discovered, as well as that predict the existence of yet newer ones. According to such theories, most new particles will turn out to be more and more massive. For heavier particles to be created, Einstein's law requires accelerators to impart more and more energy to their projectiles.

In practice, the highest energy collisions occur between the fastest moving particles. Just as in the case of two cars colliding, when the fastest moving protons collide, they create the highest energy collision. The proton projectiles used in particle accelerators are fired at speeds very close to the speed of light, the fastest possible speed in nature.

The strategy for producing the fastest possible protons consists of supplying the maximum amount of push to a proton before it hits a nucleus or another proton. Just as a discus thrower whirls around before releasing the discus, accelerators send protons around and around in giant loops, accelerating them to higher and higher speeds before they smash into their targets. To produce nearly light-speed protons, physicists build accelerator loops that are *miles* in circumference. Getting even these giant accelerators to produce fast enough protons requires a way to impart as much speed as possible to the traveling protons. Beyond that, it also requires a way to steer the protons around the loop. Superconductivity provides solutions to both of these problems.

When a moving charged particle—like a proton—travels through a magnetic field, it will veer off on a curved path. Fast particles will veer less than slow particles; strong magnetic fields will make the particles veer more than weak magnetic fields. In order to make the incredibly fast protons travel around the circular path of an accelerator, immense magnetic fields are required. As we have noted, superconducting magnets are the way to produce immense magnetic fields.

Outside of Chicago, Fermilab, the largest United States high-energy facility, has been built. It boasts a circular tunnel 4 miles long lined with superconducting magnets; these steer a proton beam along the accelerator's path. Even though Fermilab was originally designed and built with conventional electromagnets, the technological advantages of superconductivity eventually became impossible to ignore. Superconducting magnets offered designers the only practical way to achieve the 4-Tesla magnetic field (8,000 times the earth's field) needed to steer the proton beam at peak energies.

The lessons learned in building the Fermilab accelerator magnets may soon be put to use in an even larger project. The Superconducting Super Collider (SSC) project has left the preliminary drawing boards and has headed to Congress for funding. By proposing to send the protons on a path 13 times longer than the Fermilab loop and by using 60 percent more powerful superconducting magnets, designers hope for a 20-fold increase in collision energy. Priced in the neighborhood of $8 billion, the record-breaking SSC will collide protons together with enough energy to create hitherto unobtainable particles. An estimated billion dollars' worth of superconducting

Main accelerator tunnel at Fermilab showing superconducting magnets on lower ring.
(Courtesy Fermi National Accelerator Laboratory)

magnets will steer the high-energy beam of protons as they careen around the 53-mile accelerator loop. The SSC will open a window to forces and particles beyond our current reach.

Superconductivity can do more than keep protons on the curved and narrow path within the accelerator by the action of giant magnets; a different sort of superconducting device—called a *resonator cavity*—can supply the push that gives the protons their enormous speed. When we discussed maglev trains, we learned about the linear synchronous motor that uses a synchronized magnetic wave to impel the train along its track. In high-energy accelerators, scientists use a synchronized electric force to boost the speed of protons as they race around the loop.

The basic principle of a resonator cavity is quite simple. Anyone who has played on a swing set knows that the way to swing highest is to pump at just the right moment of each swing and we can only swing at one natural frequency. Pumping adds energy to the swinging motion and the only effective way to add energy is to do it at the natural frequency of the swing. Physicists call this phenomenon *resonance*. The same effect gets exploited in a resonator cavity in an accelerator.

The cavity is a specially shaped metal chamber placed in the path of the proton beam. The protons race around the accelerator loop with such tremendous speed that they pass through the cavity many times each second. To increase the energy of the protons, the cavity gives them a boost each time they zip by. Scientists know that the way to increase the speed of electrically charged particles is to use electric fields. To give the greatest boost, they use the strongest possible field.

The function of the resonator cavity is to supply as intense an electric field as possible whenever a proton passes by. It works on the same resonance principle as a swing. We feed electromagnetic waves—which are traveling electric fields—into one end of the cavity. They travel through it at the speed of light, reach the far end, and bounce back in the opposite direction. The backward wave then runs into the near end of the cavity and gets reflected once again. Now there are two waves going in the same direction: the one we are feeding in, and the doubly reflected wave. This is where the swing analogy comes in. If we use electromagnetic waves of just the right wavelength, the reflected wave will bounce off the near wall in perfect sync with the incoming wave. If this occurs, we will have doubled the strength of the wave. Just as on a swing, we can give the greatest boost by pumping at exactly the right frequency. In a resonator cavity, the trick is to let the electromagnetic wave bounce back and forth millions of times in order to produce superpowerful electric fields.

This constitutes a great idea in principle, but runs into some serious snags in practice. The big problem is that the powerful electric fields within the cavity play havoc with the electrons in the cavity's metal walls. They get swept up by intense electrical forces and create currents flowing in the walls. Such currents generate heat—by ordinary Joule heating—which robs the cavity of energy. Both the energy loss and the heat itself are bad news for the resonator cavity. The enormous electric fields in the cavity lead to destructive amounts of waste heat that have to be removed. The energy drain defeats the whole purpose of the device.

When Joule heating creates problems, superconductivity offers solutions. Modern accelerator designs turn to superconducting niobium to overcome electrical losses in resonator cavities. While resistive losses in superconductors are not precisely zero at high frequencies like they are for dc electricity, they are many *thousands* of times better than in even the best normal conductors. Electromagnetic losses in superconducting cavities are small enough that powerful electromagnetic waves can bounce back and forth enough times to build up the strength needed to propel protons in high-energy accelerators. Thus, the same giant accelerators that rely on superconducting magnets also rely on superconducting resonator cavities.

Incidentally, engineers use very similar resonator cavities for special high-accuracy clocks. Such clocks work like old-fashioned pendulum clocks, except here, time is kept by a radio wave bouncing back and forth within a superconducting cavity. The low-energy loss in superconducting cavities gives engineers a very accurate measure of time by allowing them to measure many electronic cycles of the clock.

Accelerator magnets and resonator cavities do not satiate the appetites of particle physicists for superconductivity. Once they have smashed particles together and created all manner of subatomic wreckage, they are faced with a new problem: they need to figure out what has happened as a result of the collision. Superconductivity comes in handy for this job as well.

Subatomic collision fragments are extraordinarily small and extraordinarily fleeting phenomena. Scientists have no chance of actually seeing and measuring them. Instead, they build special "bubble chambers" that allow them to observe the wake created by passing particles. A bubble chamber tracks particles in a tank full of fog. Charged particles traversing any part of the chamber literally leave a trail of fog bubbles. Scientists take roll upon roll of photographs of these fog tracks to analyze the results of atomic collisions.

Because they can't observe the particles themselves, physicists have to become detectives and deduce information from bubble chamber tracks. The way to learn the most is to subject the particles to intense magnetic fields. Just as in an accelerator loop, magnetic fields in a bubble chamber send particles off on curved trajectories. By analyzing these trajectories, physicists can piece together a detailed story of the particle collision and its aftermath. They can figure out the charge of the particles, the mass of the particles, and their lifetimes (most exotic new elementary particles lead amazingly short lives, often lasting only billionths of a second). Armed with this information, the researchers can test their theories and try to understand the machinations of the world on the most microscopic scale.

On a practical level, bubble chambers have to be surprisingly large in order to capture the microscopic violence they are designed to study. The particles go off with such speed that they cover a lot of territory very quickly and chambers need to be several feet in diameter in order to permit enough of the subatomic action to be observed. In order to bend the paths of such fast particles, very intense magnetic fields are required. Thus, bubble chambers need powerful magnetic fields over very large areas. Until the advent of superconducting magnets, bubble chambers stretched the limits of magnet performance to the breaking point and consumed huge amounts of electrical power.

The researchers of high-energy physics were among the first to recognize and commit to the potential of superconductivity to solve their problems. The world's first large magnet, and indeed the first large-scale application of superconductivity anywhere, developed from an ambitious project at Argonne National Laboratories outside of Chicago. There a group of scientists and technicians labored to build a 100-ton magnet and dewar providing the magnetic field for a 10-foot-diameter bubble chamber. Electrical power savings at least partly motivated this first large-scale project in superconductivity. Designers estimated they could save $400,000 per year from their electricity bill by investing in a superconducting magnet design. But apart from the economic benefits, superconductivity brings a level of performance to bubble chamber magnets that has rendered conventional copper electromagnets obsolete.

Monopoles and Gravity Waves

Not all the problems of elementary particle physics can be solved in giant accelerators and bubble chambers. Contemporary physics theories have predicted the existence of some phenomena that are too elusive for the standard tools of particle physicists. Researchers have therefore developed elaborate experimental techniques dedicated to the hunt for such exotic prey as magnetic monopoles and gravity waves. Superconductivity has become essential for even these specialized efforts.

Magnetic monopoles—single magnetic charges—are particles whose existence has been postulated as a way to make the mathematic description of the universe seem more balanced. The laws of electromagnetism describe a pleasing symmetry between electricity and magnetism. As we have seen, changing electric fields generate magnetic fields, and changing magnetic fields generate electric fields. Nature is filled with similar examples. However, without the existence of magnetic monopoles, a fundamental asymmetry remains between electricity and magnetism. Electrical charge comes in two forms: plus and minus. Magnets have two polarities as well: north and south. But while individual plus or minus electrical charges occur everywhere in nature, magnets always have both poles present. No one has ever observed a single magnetic pole. Permanent magnets, electromagnets, individual atoms, and even planets always have north and south magnetic poles. We can shape magnetic fields, twist the lines of magnetic force, and control the location and strength of magnets, but we can never create magnetic fields that violate the basic symmetry of north poles and south poles.

Many theoretical physicists believe that there ought to be magnetic particles that are just north poles or south poles. Such particles would impart a satisfying elegance to electromagnetic theory. Our understanding of the universe does not require there to be single magnetic poles, but the equations that describe electromagnetism would take on a perfect correspondence between electricity and magnetism if there were individual magnetic charges. So, despite the absence of any evidence for the existence of such particles, theorists have named them and worked out predictions of their properties. Do these magnetic monopoles actually exist? Experimental physicists have decided to try to find out.

According to the best theoretical guesses, monopoles are at best very scarce in the universe. They are likely to be extremely massive particles moving at nearly the speed of light. Such things are not likely to be easy to find or observe. The best chance we have to identify one is by looking for its trail. A magnetic particle with a single pole will have a magnetic signature like no other object. Thus, a sensitive magnetic detector will be the best tool to use to look for magnetic monopoles. Of course, superconductivity provides the most sensitive magnetic detectors in the world.

Scientists look for monopoles by setting superconducting traps and waiting for the monopoles to spring the traps. A superconducting monopole trap is simply a ring of superconductor. If we pass an ordinary magnet through such a ring, the north pole entering induces a current in one direction and the south pole exiting cancels that current out. The magnet leaves no trace of its presence. If a monopole passes through a superconducting ring, a current starts to flow and magnetic flux gets trapped in the loop, leaving a signature of the event.

Some theories suggest that monopoles were created in the Big Bang at the start of time and subsequently have traveled relentlessly around the cosmos in small numbers, passing effortlessly through whatever solid objects get in their way. To hunt for cosmic monopoles that hurtle into and through the earth, scientists connect together large arrays of superconducting detector rings that they continuously monitor for sudden traces of otherwise unexplainable flux.

Another theory suggests that a few monopoles may have struck the earth and become trapped by atomic forces to form a new type of molecule. To look for these trapped monopoles, scientist shovel huge amounts of earth through a large superconducting coil, looking for the characteristic signature of any monopole that may pass through the coil.

So far neither detection scheme, or any other detection scheme, has come up with conclusive proof of the existence of magnetic monopoles. Of course, no unresponsive detection experiment can ever disprove the existence of monopoles either. Hence, scientists persist in improving their superconducting traps in the hope of finding magnetic monopoles; perhaps some day they will succeed.

Monopoles are not the only missing links in the menagerie of physics. Albert Einstein predicted the existence of an entirely new phenomenon related to gravity. We all experience the effects of gravity and, since the days of Isaac Newton, we've had a pretty good understanding of its basic laws. When Einstein developed his General Theory of Relativity, the concept of gravity took on new dimensions. Einstein showed that the strong gravitational fields surrounding massive objects like stars actually distort nearby space. Such distortions can propagate as ripples in the very fabric of space, a thoroughly alien concept known as *gravity waves*.

Detecting gravity waves challenges the resourcefulness of even the most skilled experimental scientist. Unlike the familiar effects of gravity, gravity waves are incredibly weak. The only circumstances that lead to appreciable gravity waves are massive disturbances of extremely heavy bodies. Supernovas, the cataclysmic self-destruction of stars, are the kind of cosmic events that generate gravity waves. As gravity waves travel from the depths of space into our solar system, they diminish in strength to the point of nearly vanishing. The effect of a gravity wave on an object is to momentarily cause it to compress and expand, like squeezing on a rubber balloon. The effect is so weak, however, that a block of metal the size of a loaf of bread would change its length by only a billionth of the diameter of an atom. Even an object as big as the moon would only move the length of a single atom. Scientists have never before looked for such a minuscule phenomenon.

Amazingly, scientists using an arsenal of superconducting techniques are actually optimistic about their chances of detecting gravity waves. The strategy for operating a gravity wave detector consists of keeping a block of material in the (acoustically, electromagnetically, and thermally) quietest place possible, and listening intently for the appearance of any sudden unexplained disturbance. Laboratory gravity wave detectors consist of a multiton bar of aluminum carefully isolated from all seismic vibrations and cooled to liquid helium temperatures in order to reduce thermal vibrations that could otherwise mask the passage of a gravity wave.

Scientists have proposed a variety of experiments designed to detect gravity waves, and all of them exploit superconductivity in one way or another. In one approach, researchers attach a superconducting plate to one end of a massive detector bar and induce a magnetic field near the plate. Because of the Meissner effect, the superconductor pushes away the lines of magnetic force. If the bar moves even a tiny amount, the magnetic field lines will shift enough to be detected by an ultrasensitive SQUID magnetometer. The same device that monitors brain waves can therefore look for interstellar gravity waves.

Another method for detecting gravity waves uses a special kind of short-range motion radar to scan continuously the end of a massive detector. It works with the help of a superconducting microwave resonator cavity, much like the sort found in particle accelerators. In this case, the gravity wave detector gets attached to one of the resonator cavity walls. The microwave cavity acts as a very stable clock, with the time between ticks being set by the time it takes for microwaves to travel from one end of the cavity to another. Any motion by the detector changes the length of the cavity, and therefore the speed of the clock, thus giving scientists an easily monitored indicator of any sudden gravity-wave-induced movements. With carefully tuned equipment, scientists are now watching for the appearance of signals that will mark the first detection of a galactic gravity wave.

Listening to the Stars

Not everyone who gazes at the sky awaits the arrival of monopoles and gravity waves. Most astronomers find plenty already of interest among the stars and galaxies that populate the heavens. Telescopes that are directed at the visible light emanating from distant stars no longer provide our most detailed look at the universe. Radiotelescopes that sense the microwave frequencies emanating from celestial objects have become the astronomer's most powerful tool. The unequaled sensitivity and high-frequency performance of superconductive electronics has made radioastronomers the most enthusiastic supporters of the technology.

Astronomy has steered its investigations into the realm of microwave radiation because so much more can be learned from studying a broader slice of the electromagnetic spectrum than from concentrating on just the narrow band of visible light. The electromagnetic spectrum is so large and varied that we often lose sight of its unity. Radio waves, microwaves, X rays, and ordinary light waves are all manifestations of the same physical phenomenon. Electric fields and magnetic fields oscillate back and forth and propagate through space as waves traveling at the speed of light. At certain frequencies, we detect such waves with our own eyes; at other frequencies, a radio picks up the signal. Nevertheless, light and radio waves and all other forms of electromagnetic radiation are basically the same thing.

We divide up the electromagnetic spectrum according to frequency—the number of times that the electric and magnetic fields oscillate each second. Frequencies used to be quoted in units of cycles per second; we now call the unit "Hertz" (usually abbreviated as Hz), in honor of Heinrich Hertz, a pioneer in radio-wave research. The electromagnetic spectrum covers an enormous frequency range. We have learned to exploit large portions of the electromagnetic spectrum for modern technology. Radio frequencies from a megahertz (meaning one million Hertz and usually written as 1 MHz) to a gigahertz (meaning one billion Hertz and usually written as 1 GHz) are used for commercial audio and video broadcasting. Microwave frequencies from 1 to 30 GHz are used for radar, communications satellites, and

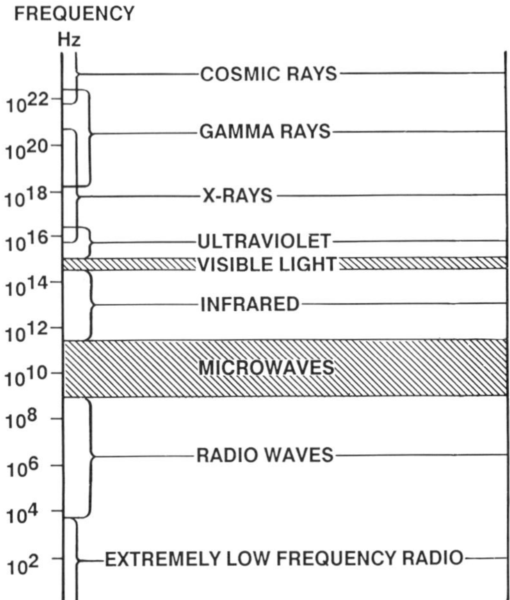

The electromagnetic frequency spectrum.

for cooking food. Above 30 GHz lie millimeter and submillimeter waves. Finally, the electromagnetic spectrum enters ranges that we associate with light and radiation more than radio waves. Infrared, optical, ultra-violet, X rays, and gamma rays fill out the remainder of the electromagnetic chart.

To date the most conspicuous gap in our use of the electromagnetic spectrum has been in the 30 GHz to several THz (terahertz: a trillion Hertz) region. Commercial applications for these frequencies have been very sparse—not because we have run out of useful things to do, but because the technical challenges of using such high frequencies have seemed insurmountable. Superconductivity promises to change technology's attitude toward this part of the electromagnetic spectrum.

Superconductivity has two things to offer for high-frequency electronics: high sensitivity and wide operating frequency range. Because of these attributes, superconductivity is opening up new opportunities to exploit the high-frequency portion of the electromagnetic spectrum. With superconductive electronics, crucial circuit elements like mixers and amplifiers can be built to operate at previously unusable frequencies.

Radio and television waves travel through space at frequencies of millions to trillions of Hertz. How do we make use of frequencies this high to broadcast voices and music? Even if a super set of speakers could change the electromagnetic waves directly into sound waves, our ears could never hear frequencies so high. Instead, the broadcast frequencies must be converted into audible frequencies by an electronic device known as a mixer. Mixers convert signals at high frequencies down to lower, more usable frequencies. Think of a mixer as an electronic device that can convert a soprano into a baritone. The action of a mixer is known as "downconverting," and virtually any nonlinear device can do the job. As we may recall, nonlinear devices are ones that don't have straight-line current–voltage characteristics. Semiconductor devices like diodes and transistors are nonlinear devices, and they are commonly used as mixers. As we also recall, superconducting tunnel junctions are nonlinear devices, and they too can be used as mixers.

Like all electronic devices, mixers are never perfect. When a mixer converts the frequency of an incoming signal, it inevitably adds a little noise to the output signal as well. Played on a radio, such mixer noise sounds like a monotonous hissing heard in the background. On a television set, mixer noise appears as snow. The noise level determines the ultimate sensitivity of any type of receiver. Sensitivity means the ability to receive tiny signals. Too much noise will blot out the smallest signals. For television, the amount of noise determines the amount of snow on the screen. For radar, noise determines the range over which an airplane can be detected. For radioastronomy, noise sets the limit on how far into the cosmos we can observe. A key advantage that superconducting mixers offer is their almost immeasurably small noise levels. When sensitivity becomes the overriding issue in any application, superconducting mixers are unsurpassed.

Of all the users of advanced electronics, radioastronomers have the greatest need for sensitive, low-noise devices that can operate over a broad frequency range. Their work requires them to try to capture extremely weak signals originating from across the galaxy and beyond. To aid their cause, radioastronomers live in a very quiet world. They locate their "telescopes," actually big dish-shaped antennas, in isolated locations to get away from city noise. The antennas sit on arid mountain tops so that water vapor in the air does not absorb the signals they seek. The radioastronomers turn their dishes towards deep space, where sometimes the only signal to be found is the faint, ubiquitous rumbling left over from the primordial "big bang."

A plethora of faint radio signals emanating from gas molecules emerge from interstellar space. Like tiny tuning forks, all molecules have natural ringing frequencies. Collisions with other molecules or with photons of light start the molecules ringing. Instead of emitting sound, molecules give off radio waves at characteristic frequencies. These waves form a fingerprint of emissions that uniquely identify the ringing molecules. Although we think of space as a perfect vacuum, thin clouds of molecules permeate the space between stars. The vacuum of space is polluted with such gases as carbon monoxide, OH (a fragment of water, H_2O), ammonia, cyanide, and even ethyl alcohol. Radioastronomers have even detected a number of space chemicals that are far too unstable to exist on the earth.

Where do all these extraterrestrial pollutants come from? They are certainly not the tailpipe exhaust from flying saucer thoroughfares. As stars evolve and grow older, they reach a point where they run out of nuclear fuel. When this happens, violent explosions can cast off part of the stellar matter. The clouds drifting between the stars are remnants of these ancient suns. Scientists carefully study interstellar molecules because these atoms give us clues about the birth, life, and death of star systems, including our own.

In some ways, we know more about the chemistry near the Orion nebula, many light-years away from earth, than we do about our own upper atmosphere. Each observed emission frequency identifies a chemical present in the interstellar cloud. Radioastronomers deduce quantity, temperature, and even speed of these cosmic chemicals from their radio signatures.

Besides these discrete chemical emission lines, radioastronomers also hunt for a more diffuse cosmic signal: the all-pervasive cosmic background radiation from the "big bang" origin of the universe. According to present theory, the universe is still expanding and cooling from a primordial fireball from which all matter and energy emanated billions of years ago. Light and radio waves from the initial "big bang" have been echoing and reverberating through the cosmos ever since. Today this radiation manifests itself as a microwave signal covering a wide frequency range and appearing to come from everywhere in outer space. While part of this radiation can be detected at low microwave frequencies, much of the power propagates at higher frequencies, where superconductivity can provide the best detectors. Astrophysicists investigate this cosmic signal to quantitatively test the "big bang" theory and other cosmic models. Time travel may be an impossibility, but the cosmic radiation left over from the "big bang" allows astronomers to see backwards into the past all the way to the origin of the universe.

To get their all-important cosmic data, radioastronomers point their giant telescope dishes to these cosmic radio sources and listen carefully. The quality of their data, the size of the detectable signals, and the range that they can scan in the heavens all depend on the noise levels of the radiotelescope receivers. Radioastronomers therefore go to great lengths to build the quietest possible receivers.

Superconducting mixers are beginning to supplant their semiconducting cousins in radiotelescope receivers operating above 100 GHz. The Owens Valley Radiotelescope, operated by the California Institute of Technology, has used superconducting receivers since 1983. The National Radio Astronomy Observatory, and observatories in Japan, Germany, the Netherlands, Australia, and a number of other countries are building superconducting mixers for radioastronomy as well. Optical telescopes require high-quality mirrors and lenses for their ultimate performance; radiotelescopes require superconducting mixers.

Super Television

Although radioastronomy currently constitutes the dominant application of the above-30-GHz region of the electromagnetic spectrum, this near-monopoly is not expected to last much longer. The advances in high-frequency electronics brought forth by superconductivity promise to open huge new areas of practical endeavor.

There are a number of reasons why engineers are eager to exploit higher frequency regions of the electromagnetic spectrum. In the broadcasting industry, a competition for scarce resources drives the interest in higher frequencies. Currently, a fairly narrow slice of the electromagnetic spectrum serves the requirements of television and radio broadcasters, radar operators, and telephone communication companies. As more and more people are vying for the same airwaves,

Owens Valley Radiotelescope, operating with a superconducting mixer.
(Courtesy Caltech)

conflicts are inevitably arising. The competition for precious operating frequencies has become so fierce that commercial broadcasters have to fight off competitors in order to keep their licenses and international congresses have to convene in order to settle disputes about broadcast frequencies.

Superconductivity's ability to open up higher frequency ranges for electronics applications may drastically affect the commercial broadcast industry. The evolution of the television industry has been locked into the abilities of the electronics industry. Television broadcasting takes place at frequencies under 1 GHz because early receivers could not be built to receive higher frequency signals. Coast-to-coast transmissions between studios became possible with the development of satellites and higher frequency electronics. As microwave technology has improved and dropped in price, it has become possible for consumers with enough money to eavesdrop on satellite transmissions as well. Although the satellite-to-consumer market continues to grow, its growth is limited by such drawbacks as the requirements for large dish antennas and costly electronics.

Many of the problems of satellite-to-home broadcasting result from the operating frequency: about 8 GHz. Once the operating frequency is set, many features of the broadcasting system are determined. Antenna dishes must be several meters across in order to receive enough power from the orbiting satellites. The number of channels is limited to about a hundred by the available frequencies allotted. Even the number of satellites orbiting the earth becomes regulated by the need to avoid interference.

If superconductive electronics opens up a portion of the millimeter-wave spectrum for commercial broadcasts, so that a 10-fold increase in frequency occurs, satellite television will completely change. The antenna dishes could then shrink to the size of dinner plates with no degradation in performance. Such antennas would neither be eyesores on urban rooftops or strains on urban budgets. If channels were allotted the same amount of space at the new frequencies, there would be room for ten times as many at ten times the frequency. Thus a single satellite could beam a thousand simultaneous channels down to the earth. In addition, many more satellites could be squeezed into earth orbit without interference because of the decreased wavelength of the transmissions. As a result, tens of thousands of channels could become available to the consumer.

The possibility of transmitting so much information would revolutionize the broadcast industry in ways we can only speculate about. Certainly not all the channels would be devoted to video transmissions. Newspapers and magazines may forego the costs and delays of printing and delivery by directly transmitting copy to their subscribers' receivers. Telephone companies could reserve some part of the frequency space for transoceanic connections, reducing international toll charges to the level of interstate rates. If superconductivity enables the commercial development of millimeter-wave electronics, the entire communications industry will undergo the biggest revolution yet.

We have explored a rather diverse set of superconducting applications being pursued by scientists and engineers from a number of fields. Some of the most developed applications of superconductivity are represented here, as are some of the most obscure. Whereas the largest projects in science incorporate superconductor technology, small-scale science depends on superconductors as well. Superconducting circuits and devices are now routinely used in hundreds of laboratories around the world for performing a variety of experiments and for making precise measurements. Although very few of these applications generate as much excitement as the more commercial uses of superconductivity, superconductors have always earned their keep in the business of science.

CHAPTER 12

The Superconductivity Business

We have explored superconductivity as a science and as a technology, but superconductivity is also today a growing business. The tremendous publicity associated with recent laboratory breakthroughs in superconducting materials should not completely overshadow the fact that applied superconductivity supports a sizable worldwide industry.

The superconductivity industry accounts for nearly $400 million in annual revenues for products and research activities. Of course, on the scale of a trillion-dollar U.S. economy, superconductivity constitutes only a tiny industry. However, we should note that the $400-million figure includes only the superconducting components of larger instruments and machinery that make use of superconductivity. From this standpoint, superconductivity already forms part of a billion-dollar economy. So, while it may not be not the biggest technology business around, it should by no means be considered a mere laboratory curiosity. While the economic impact of the new high-temperature superconducting materials may be difficult to estimate, the superconductivity industry anticipates continuing growth leading to multibillion-dollar revenues by the turn of the twenty-first century.

The element niobium has been the breadwinner in the superconductivity business. Its alloys support the superconducting magnet and wire industry; niobium itself and niobium nitride dominate superconductive electronics. But apart from the unifying role of niobium in the majority of superconductor applications, the superconductivity industry embraces a diversity of enterprises.

Today's superconductivity business centers around four general activities: the production of superconducting magnets for small-scale and large-scale applications; the manufacture of superconducting rods, wire, and cable; the fabrication of superconductive electronic devices and circuits; and research and development activities on superconducting materials and devices. Even before the recent surge of interest in superconductivity, all these areas experienced continuing growth.

Selling Superconducting Magnets

In terms of sheer economic impact, niobium-alloy superconducting magnets dominate the superconductivity industry, accounting for more than half of its total revenues. Furthermore, the lion's share of the superconducting magnet business is devoted to supplying magnets for magnetic resonance imaging machines. Several hundred of these advanced medical-diagnostic instruments are being sold to hospitals each year and the market continues to grow. The leading companies in the magnetic resonance imaging field are Siemans in Germany, and General Electric, whose medical instrument manufacturing facilities are located in Britain and in the United States. Magnetic resonance imaging (MRI) machines sell for about $3 million each, with the superconducting magnet accounting for about 10 percent of the price. The $150 million in magnet sales for MRI machines accounts for about 75 percent of the total niobium-titanium wire business. Current predictions foresee annual MRI magnet sales in excess of $400 million within the next few years. This application alone guarantees the continued health of the superconducting magnet and wire businesses.

The high-energy physics community constitutes the second major customer for large superconducting magnets. Giant particle accelerators are not built every day, but when they are, modern designs now call for superconducting magnets. The first operational superconducting accelerator—known as the Tevatron—came on line in 1983 at Fermilab, near Chicago. The project took 4 years to complete and includes 1,000 superconducting magnets. A second big superconducting accelerator project, known as the Hadron-Electron Ring Anordnung (HERA), is currently under construction in Hamburg, Germany. Other large-scale physics machines for the high-energy physics community and the plasma physics community, respectively, include the Rela-

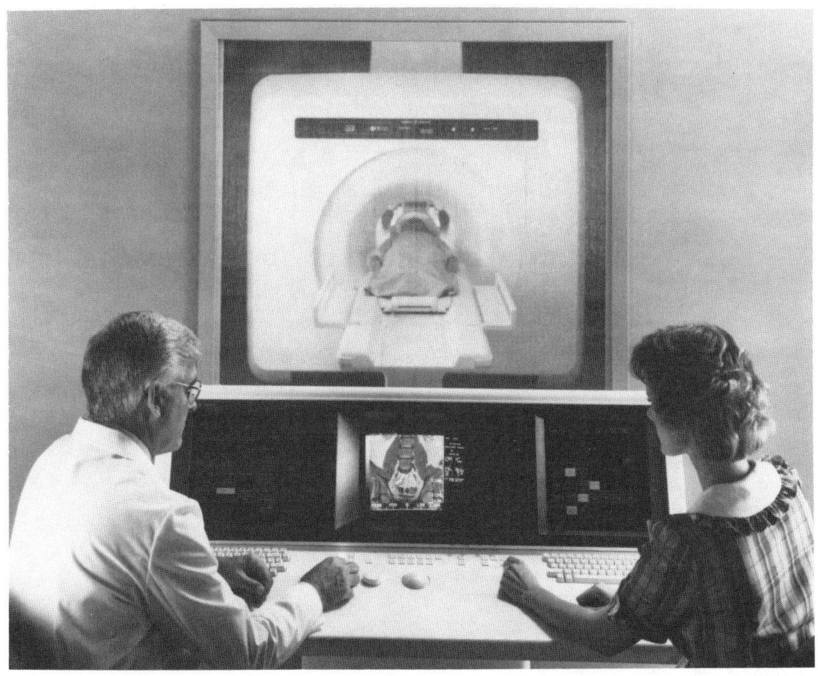

Magnetic Resonance Imaging system. (Courtesy General Electric)

tivistic Heavy-Ion Collider, and the International Large Coil Fusion Project at Oak Ridge National Laboratory in Tennessee.

Looming on the horizon for the high-energy physics community is the Superconducting Super Collider (SSC), which, if built, would be by far the largest particle accelerator ever constructed. The Super Collider project represents the most ambitious application of superconducting magnets ever proposed. The present design calls for 10,000 superconducting magnets along a 53-mile loop, the construction of which would stimulate a lengthy boom period for the superconducting magnet industry. A fierce competition between many proposed sites will culminate in a choice for a location for the massive high-energy physics facility. Some 4,500 workers will be needed for the construction project and the facility will generate 2,500 permanent jobs. Many states have been vying to become the site of the Collider, and many companies are vying for the lucrative superconducting magnet contracts for the project.

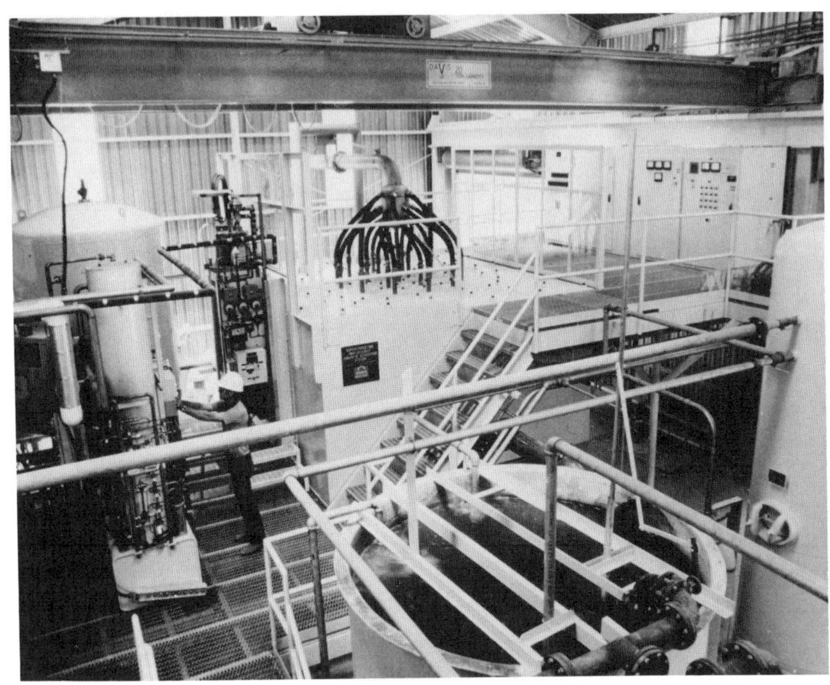

Superconducting magnetic separator at the J. B. Huber clay-processing plant in Wrens, Georgia. (Courtesy Eriez Magnetics)

Powerful superconducting magnets can also be used for separating magnetic materials from ores or seawater. One company, Eriez Magnetics, has designed a specialized system for this purpose and in 1986 sold a prototype to the J. B. Huber Corporation. Huber runs a clay-processing plant in Georgia and uses the superconducting separator to purify clay for use in ceramics, paints, and paper coatings. After 1 year of operation, Huber reported energy savings of $150,000 compared with the operation of more conventional purification equipment and therefore plans to buy an additional superconducting separator system in the future. Eriez expects that its system will successfully compete with existing technology for magnetic separators for metal ores, seawater minerals, and other applications throughout the industrial marketplace.

Although large-scale magnets dominate the superconducting magnet business, a thriving market exists for small magnets designed for scientific instruments and experiments. Most of these magnets are used in physics and engineering laboratories, but research is underway to develop small superconducting magnets for specialized commercial instruments. One major supplier of low-temperature instrumentation, the British firm of Oxford Instruments, plans to use superconducting magnets in advanced X-ray lithography machines that produce microscopic circuit patterns for the integrated circuit industry.

Oxford has also received a contract from IBM to build a synchrotron using superconducting magnets by 1990 that IBM will use for the fabrication of advanced integrated-circuit chips. Synchrotrons are high-energy electron accelerators that have been used by particle physicists for many years. IBM's interest occurs because synchrotrons produce an intense electron beam that can also be used for patterning electronic circuits on an extremely microscopic scale. Conventional synchrotrons occupy about 10,000 square feet of floor space; Oxford's compact superconducting synchrotron will cover a 6-foot by 15-foot area. Such minisynchrotrons may become a mainstay of the electronics industry in the future; several Japanese high-technology companies have already announced their intention to market such devices in the 1990s.

For all of these specialized applications, the liquid helium cooling requirements of niobium-alloy superconductors represent acceptable costs for the required performance. Independent of the development of new high-temperature superconductors, the superconducting magnet industry based on niobium-alloy superconductors will continue to grow in the years ahead.

The Business of Wire

The manufacturers of superconducting magnets are not the same companies that produce superconducting wire. A separate industry supports each function in the superconducting magnet business. For example, niobium-alloy manufacturers support the superconducting magnet business with the fabrication of niobium-titanium rod that can be drawn into wire, which, in turn, can be fashioned into cable. In the United States, the Wah Chang subsidiary of Teledyne produces most of the niobium-titanium rod that goes into superconducting wire. Several other companies then produce multifilamentary superconducting wire from this kind of rod. Important suppliers of superconducting wire include Supercon and Intermagnetics General in the United States, Oxford Instruments in Great Britain, and Hitachi in Japan.

Wire manufacturers have steadily improved the performance of niobium-titanium alloy wire since the 1960s. Commercial superconducting wire generally consists of many filaments of niobium-titanium alloy imbedded in a copper matrix. Multiple filaments in conjunction with cleverly engineered twist geometries lick the problems of transporting alternating current electricity through the composite copper–superconductor wires. Wires containing many thousands of superconducting filaments offer the highest current-carrying capability. High-current commercial wire may contain up to 4,164 filaments; experimental wires have been tested with as many as 40,000 filaments. The heaviest versions of commercial superconducting wire are designed to handle up to several thousand amperes of current. The thinnest wire—measuring only three-thousandths of an inch in diameter—can still easily carry a few amperes. But even with such performance, wire makers cannot rest upon their laurels. Superconducting wire continues to be improved and developed further for increasingly demanding applications.

Industry insiders estimate that the superconducting wire business generates annual revenues of $15 million, most of which comes from sales of niobium-titanium alloy. Superconducting-wire manufacturing has been an established enterprise for quite some time, with firms like Supercon doing business since the 1960s. In order to remain competitive, companies in the business must pursue vigorous research programs in new materials and techniques.

In recent years, there has been a push to bring other superconductors into the stage of commercial development. Engineers have developed processes for producing wire from some of the A15 superconductors we talked about earlier. Materials like vanadium-gallium and niobium-tin remain superconducting in higher magnetic fields and at higher temperatures than the niobium-titanium workhorse material. These materials require much more elaborate processing in order to produce usable wire, but engineers have made significant progress in their development. For a while, it appeared that niobium-tin wire would be used in the Super Collider magnets, but designers eventually decided to go with the more established niobium-titanium technology. Nonetheless, the demand for stronger magnets for the next generation of magnetic resonance imaging machines and other magnet-based systems strongly motivates the continued development of advanced superconducting wire. In addition, all the superconducting-wire manufacturers are also actively pursuing the development of wire made from the new high-temperature superconductors.

A much smaller support industry exists for the purpose of supplying superconducting raw materials for thin-film circuit technology. For example, a number of companies produce specially machined sputtering targets of niobium and other materials that are used in the fabrication of superconducting films. These same companies have also become suppliers of raw materials for the deposition of high-temperature superconductor films.

Junctions for Sale

Turning our attention to thin-film superconductor applications, we find that the superconductive electronics business falls under two general headings: product-oriented efforts and research-oriented efforts. Taken together, the efforts in superconductive electronics comprised a $40-million-a-year industry before the discovery of high-temperature superconductivity. The dollar amount may be small, but the superconductive electronics industry has only reached its infancy as a commercial enterprise.

The veteran firm involved in superconductive electronics is Biomagnetic Technologies, which began operations under the name of S.H.E. (Superconductivity, Helium, and Electronics) back in 1970. For more than a decade, S.H.E. pursued the business of manufacturing commercial SQUID systems for the scientific community, SQUIDs that are used by physics and engineering laboratories around the world as sensitive magnetometers and electrical measurement systems. Commercial SQUIDs started out as devices that were machined from solid blocks of niobium; the superconducting loop and weak-link structure that form the SQUID were made from carefully crafted niobium parts. By the 1980s, these so-called bulk SQUIDs were replaced by thin-film niobium and lead-alloy circuits. Such SQUIDs can be purchased from Biomagnetic Technologies (which uses the name BTi in the marketplace) and other companies today.

By 1985, BTi decided to concentrate its efforts upon one specific application area in SQUID electronics: biomagnetic sensing. The tremendous sensitivity of SQUID magnetometers would be applied to the technology of medical diagnosis. The company now produces a SQUID-based product called the *neuromagnetometer*, an instrument designed for the noninvasive examination, measurement, and precise location of brain activity. The device has received Food and Drug Administration approval as a clinical instrument and by 1988 had already been installed in more than 50 medical research facilities.

The neuromagnetometer may well become the diagnostic instrument of choice for epilepsy and would therefore become an essential tool for major hospitals everywhere. At $1 million per system, such expanded use promises a sizable economic base for SQUID magnetometry. Future plans at BTi call for more advanced versions of the instrument as well as for other SQUID-based diagnostic devices. BTi is banking on the idea that biomedical instruments using SQUID electronics will become a standard part of the modern physician's diagnostic arsenal.

Other companies are concentrating on different SQUID applications. Quantum Design, a company started by former S.H.E. researchers, has taken up where S.H.E. left off in its original product line. The original S.H.E. marketed a state-of-the-art SQUID susceptometer, a device for making ultrasensitive measurements of the magnetic properties of materials. Quantum Design now markets the next-generation version of this device under the name of the "Magnetic Property Measurement System." The instrument sells for over $100,000 and demand for it continues to outstrip Quantum Design's capacity to build it. Research labs and universities often wait for more than a year to acquire one of these coveted scientific instruments.

Quantum Design recently formed a subsidiary—Quantum Magneduction—in order to market a SQUID instrument designed to detect corrosion in metal. When rust forms on metals, SQUIDs can detect subtle changes in the magnetic properties of the metal. Such a system holds great interest for industrial firms that must inspect extensive metal installations for corrosion. Petroleum pipelines are a prime example.

Other firms also provide various SQUID instruments. Oxford Instruments currently markets SQUID electronics for scientific customers, but has plans to enter the biomagnetic sensing field as well. A number of other companies are also in the SQUID business, but are not involved in the manufacturing of commercial products. We come back to this part of the SQUID market shortly.

Although the discovery of Josephson tunneling predated the discovery of the SQUID, SQUID electronics established a commercial beachhead long before Josephson junction electronics. SQUIDs could function with crude Josephson junctions or with easy-to-make weak-link structures, which enabled commercial products to be built quite early on. In contrast, the technological problems of making reproducible, controllable Josephson junctions presented a major obstacle to the early deployment of Josephson electronics. IBM tackled the problems of Josephson electronics head-on with its program to build a Josephson computer. The IBM Josephson computer project was certainly aimed toward the goal of a real product, but the project ended long before the goal was reached. The IBM effort did not include any smaller and less-ambitious intermediate products along the way; a complete computer was to be the output of the development program. When IBM closed down its Josephson computer development program in 1983, many people thought that the end had come for superconductive electronics technology. But despite the decision at IBM, other researchers felt that there existed a number of unexploited market niches for more readily realizable Josephson electronics products.

One such scientist was Sadeg Faris, who had spent 8 years working in the IBM Josephson program. Shortly before the demise of the computer project, Faris left IBM to start a new company called Hypres. The plan was to establish the credibility of Josephson junction electronics technology by developing a series of small-scale, high-performance electronics products. As we have noted, Hypres would try to demonstrate that Josephson electronics-based instruments could be brought to the marketplace in the form of reliable, state-of-the-art, specialized instruments. Venture capitalists were willing to bet that, by adopting this strategy, Hypres could succeed where IBM had failed. If it could be done, then Josephson electronics technology would finally be ready to tackle more ambitious tasks like superconducting computers and other large-scale electronic systems.

Hypres's first product is a high-speed laboratory instrument designed for advanced electronic test measurements. Hypres provides a complete, self-contained system—known as the PSP-1000 Picosecond Signal Processor—to its customers. The PSP-1000 includes a niobium thin-film Josephson electronics integrated circuit, a considerable amount of sophisticated conventional electronics, and a novel self-contained liquid helium cooling system for the supercon-

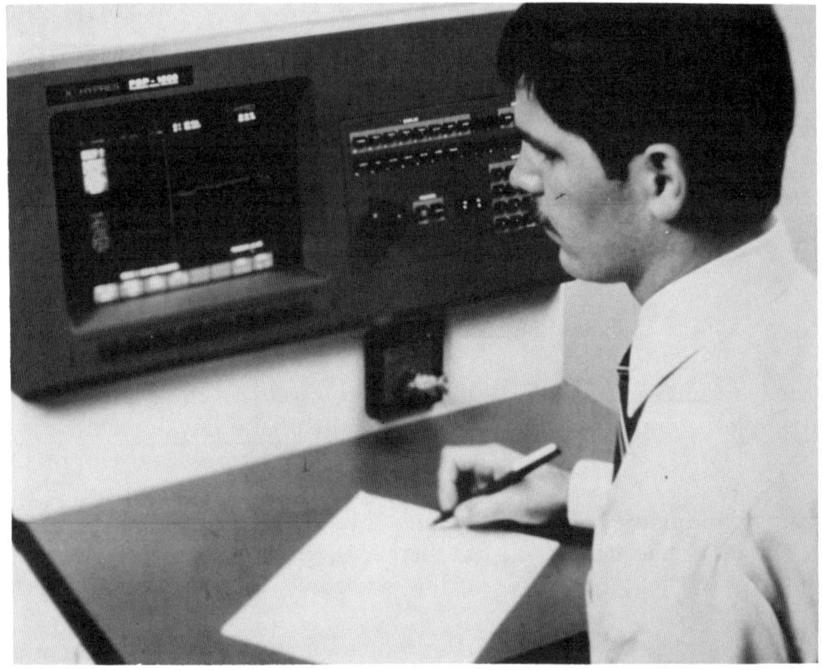

Picosecond signal processor. (Courtesy Hypres)

ducting circuitry. The instrument can function as the world's fastest oscilloscope, a device used to measure electrical signals that vary in time. In this mode, the PSP-1000 can accurately track signals that last for only 5 trillionths of a second (5 picoseconds). Such performance beats the nearest competitors by a factor of five.

Hypres began to deliver its first product in 1987, thus marking the beginnings of true commercialization of Josephson junction electronics. The basic system sells for over $120,000 and, according to Hypres, high-tech customers are lining up for the instrument. The company plans to market other Josephson electronics-based products in the future. In the sales literature from Hypres, superconductive electronics is hailed as "the third age in electronics technology," following in the footsteps of vacuum tube and transistor technology. The next decade may determine the truth of this claim. Meanwhile, the dream of Sadeg Faris, in keeping with his previous work, remains to develop and market a Josephson junction computer. If Hypres succeeds with its smaller scale superconducting products, he may someday get his chance.

Research: The Business of Tomorrow

Commercial products like magnetic resonance imager magnets and SQUID instruments represent a large piece of the economic pie in applied superconductivity, but a major fraction of the actual work done in the field comes under the heading of research and development. Many superconducting applications still have to prove their worth in test programs before anyone will foot the bill for commercial products. Whether done under the auspices of the government or as part of the efforts of private industry, the frontiers of superconducting technology continue to expand in the research environment.

As we have seen, only a few specific applications of superconductive electronics technology have made it to the commercial development stage. Many others still await the identification of a market or demonstration of sufficient performance at an acceptable cost to warrant industrial investment. Thus it has been the United States Department of Defense (particularly the Office of Naval Research and several Air Force agencies), NASA, the National Science Foundation, and several other agencies that have actively supported superconductive electronics research since the early days of the field. Government interest spans a variety of application areas including communications, radar, surveillance, signal processing, and other aspects of high-performance electronics for ground-based and satellite systems. Since superconductive electronics is smaller, lighter, and less power hungry than its competitors, it continues to provide a special attraction for spacecraft engineers.

Superconducting applications in high-frequency electronics are being developed at a number of laboratories including government and academic facilities like the National Bureau of Standards, Lincoln Labs, the Jet Propulsion Laboratory, and the University of California at Berkeley, as well as at industrial laboratories like TRW. Research programs at these laboratories develop high-frequency amplifiers, oscillators, and other electronic components for advanced communications systems of the future. These electronic components are being developed at the expense of NASA and the Defense Department at academic and industrial labs, but they may someday find their way into a broad spectrum of commercial applications. High-frequency analog electronics is expected to play a growing role in a variety of future technologies.

Digital electronics receives considerable attention from government funders as well. Although no program exists in the United States for the construction of a Josephson electronics computer, many research efforts are underway to develop various high-performance digital circuits based on superconductive electronics. Laboratories developing high-speed, high-precision signal-processing circuits include Hypres, TRW, and Westinghouse. In the United States, these laboratories have assumed the leadership role in the further development of methods for fabricating superconductive electronics.

In Japan, more ambitious digital applications continue to dominate the research community. The government's Ministry of International Trade and Industry (MITI) has, throughout the 1980s, been coordinating an effort to develop a Josephson computer and many of Japan's largest technology firms have been involved. Fujitzu, Hitachi, Nippon Electric Corporation, Nippon Telegraph and Telephone, and the government-owned Electro-Technical Laboratory have all been involved in the computer project. Outgrowths of this effort have been the development of high-quality niobium Josephson junctions and advanced integrated circuit techniques that are now used worldwide.

Superconducting analog and digital electronics combine thin-film SQUIDs and other Josephson junction circuits with microwave components, integrated-circuit techniques, and other advanced electronics technologies. In advanced superconductive electronics, SQUIDs are no longer used only as sensors for magnetic fields but rather are exploited as high-speed switching devices, storage devices for electrical currents, and as specialized components in high-frequency circuits. Ultrahigh-performance superconductive analog-to-digital converters, for example, operate by ingenious applications of SQUID technology. Thus there are many customers for SQUIDs beyond those involved with magnetometry. On the other hand, magnetometry is not the only sensor application in superconductivity.

Although research on high-frequency analog electronics and high-speed digital electronics has continued to expand, the area of superconducting sensors has experienced the most rapid growth. Sensors are devices that allow us to detect and measure various kinds of signals—magnetic fields, electrical current, temperature, light, and others—and superconducting sensors can often measure weaker signals than any other kind of device can detect.

The limits of sensor performance are set by the intrinsic sensitivity of the device—how small a signal it can detect—and by the presence of noise. Noise, whether of internal or external origin—will obscure the smallest signals regardless of the sensitivity of the sensor. Superconducting sensors deliver high performance in both sensitivity and noise immunity. Superconducting sensors come in a number of forms, with SQUIDs being only the most familiar. Granular superconducting films—thin films composed of small grains of superconductor imbedded in an insulator—have demonstrated high sensitivity to various kinds of radiation. Josephson junctions can also be used as sensitive radiation detectors, as we discovered when we explored superconductivity's role in radioastronomy.

In general, who needs high-performance sensors? In almost every field of endeavor, ultrasensitive measurements play an increasingly important role. Detecting small signals of various kinds has become essential in many technological pursuits. In industry, for example, the increased use of robotics demands the utilization of sensors for guidance, flaw detection, and critical monitoring of temperature, position, and a host of other elements in manufacturing processes. In medicine, as we have already seen, superconducting sensors can monitor brain activity, measure iron content in the liver, and search for hidden tumors within the body. In surveillance, superconducting sensors may find their way into sophisticated infrared, millimeter-wave, and even submillimeter-wave imaging systems. Research on superconducting sensor applications has already become a $20 million a year effort and may lead to what some experts expect to be a $1 billion a year market by the mid-1990s.

Superconductive electronics research has entered a period of rapid growth. Before the discovery of high-temperature superconductivity, superconducting electronics research in the United States employed perhaps 200 researchers. Apart from product-oriented firms like Hypres and BTi, most superconductive electronics research groups at industrial, government, and university laboratories consisted of 20 people or less. But in 1987, the growth in all areas of superconductor research was explosive. People were moving in and out of the field in large numbers as many companies new to superconductivity tested the waters of the technology. It will take at least several years to reestablish a stable size for the superconductive electronics research community.

Superconductivity around the Globe

Electronics is not the only area of applied superconductivity research. Whereas superconducting magnets have become a viable product, other large-scale applications have not. Most of the energy and transportation applications of superconductivity we have discussed in this book have not reached any commercial status, but continue to be developed in the form of test and prototype systems in research projects. Nonetheless, some of these projects are quite substantial in scope and have considerable impact both in terms of jobs and revenues for the contractors involved.

In the electrical power industry, Superconducting Magnetic Energy Storage (SMES) is being developed under a United States Department of Defense multiyear contract. Two competing teams—led by EBASCO Services, Inc., and Bechtel National, Inc.—are working on the conceptual design and the selection of the components, materials, and site for the project. The proposed system will be a football-field-sized superconducting storage coil capable of storing 30 megawatt-hours of electrical energy, a day's worth of power for a thousand homes.

In Japan, the Central Research Institute of the Electric Power Industry has earmarked $250 million to be spent on work related to superconductivity over the next decade. The majority of these funds will be spent in developing applications for existing superconducting technology. Only unexpectedly rapid progress in the new high-temperature superconductors would divert a greater share of the money away from the existing materials. In the United States, the Electric Power Research Institute also anticipates increased funding for superconductivity research.

As far as transportation applications for superconductivity are concerned, Japan Railways General Institute continues to lead the way in the development of maglev train systems. The Magnetic Levitation Railway System program has been operating with a $5-million annual budget; with the discovery of high-temperature superconductivity, this budget has been doubled. In the United States, there is renewed interest in maglev technology. A company called American

Mag-Lev has proposed the development of a superconducting rail line running from Kennedy Airport through Manhattan down to the Delaware Water Gap in the Poconos. And in terms of personal transportation systems, both the Ford Motor Company and General Motors have research programs in progress for evaluating the future role of superconductivity.

While industrial firms and consortia are becoming increasingly involved in superconductivity research, governments around the world continue to play the major role in the technological development of superconductivity. The United States government has recently been spending some $50 million a year on superconductivity research and development efforts. As a result of the discovery of high-temperature superconductivity, these amounts are expected to substantially increase. An initiative by the Reagan administration calls for tripling the government's spending on superconducting research. The government operates its own major research laboratories out of both the Department of Energy and the Department of Defense, as well as supporting research in private industry and at universities. In the wake of the discovery of high-temperature superconductivity, major new funding programs from the Defense Advanced Research Project Administration (DARPA), the Department of Energy, and other agencies are coming into existence. The DARPA program alone will administer an annual budget in the tens of millions of dollars.

In Japan, substantial research and development efforts are underway in a number of key areas of applied superconductivity. For many years, the Ministry of International Trade and Industry (MITI) has played a central role in the development of superconducting technology. In the area of electrical power, for example, MITI has organized a consortium of 11 firms to work on a $250-million, 8-year project for the development of a 200,000-kilowatt superconducting generator. As mentioned earlier, MITI has been the driving force behind Japan's Josephson computer effort as well.

By 1988, well over a hundred companies in Japan employed over 1500 scientists in various areas of applied superconductivity research. Major industrial firms were engaged in development efforts in large-scale superconducting machines, high-performance superconductive electronics, and advanced superconducting materials. Among the larger programs in Japan are wire and cable development by Sumitomo Electrical Industries, a diversity of programs at Sanyo, and the large-scale computer development effort going on under the auspices of MITI. A new industrial association for the development of superconductor technology in Japan is expected to include over 100 companies. The association may assume an international character and become open to foreign firms as well.

Although the United States and Japan constitute the largest centers for superconductivity research, many other nations pursue the technology as well. In Canada, for example, the government's Industrial Research Assistance Program is organizing a consortium of Canadian industrial and university laboratories to develop high-temperature superconductors. Great Britain has been involved in superconductivity research since the 1930s and currently boasts such active industrial firms as Oxford Instruments and Plessey, as well as long-established research groups at Cambridge, Oxford, and other universities. The British government sponsors superconductivity research through its Science and Engineering Research Council, as well as through the Department of Trade and Industry.

On the European continent, major industrial concerns like Phillips in the Netherlands and Siemens in Germany are active in superconductor technology. Substantial government programs in superconductivity have existed in the Soviet Union for many years. Of particular emphasis have been long-term programs in superconducting power generation, transmission, and storage, magnetic fusion confinement, and superconductive electronics. With the advent of high-temperature superconductivity, Soviet superconductivity research has experienced the same explosive growth that has occurred elsewhere.

The existing business of superconductivity—as distinct from the potential business that may eventually develop out of the discovery of high-temperature superconductivity—has grown up with another enterprise in its shadows. Everything accomplished to date by the commercial superconductivity business has been inseparably linked to the helium business. About 1.5 billion cubic feet of liquid helium is produced in the United States each year, with about 25 percent earmarked for superconducting applications. Most of the world's helium supply comes from natural gas wells in the United States, with only about 5 percent coming from European wells and elsewhere. The world's largest supplier of liquid helium is Air Products and Chemicals, which also dominates the liquid nitrogen market. Air Products' liquid helium business accounts for about $25 million in annual sales.

The largest growth area has been helium for magnetic resonance imaging machines, although engineering advances have made the latest machines increasingly efficient in their use of liquid helium.

The superconductivity industry—particularly the research community—also comprises a major market for helium liquefaction equipment. The largest supplier of this type of equipment is Koch Process Systems, whose helium liquefier business accounts for sales of about $6 million a year. The same firm also produces nitrogen liquefiers, which may become a product in greater demand as high-temperature superconductors are developed for commercialization. Until such time as commercial products can be produced using high-temperature superconductors, the presence of liquid helium will remain a superconductor fact of life.

The potential for applied superconductivity casting aside its dependence upon liquid helium has been the biggest development ever for the field. Throughout this chapter we have repeatedly made reference to the discovery of high-temperature superconductivity. One cannot avoid referring to the recent breakthroughs in superconducting materials in any discussion of superconductivity, particularly when the topic is the economic status of the technology. Unquestionably no single event in the history of superconductivity has more far-reaching consequences for the future status and growth of the superconductivity industry.

During the three decades in which applied superconductivity has made its first appearance in the commercial world, two issues have been dominant: the technical feasibility of superconducting applications and the economic feasibility of helium-cooled systems. It would be a great exaggeration to say that researchers have overcome all the technical obstacles standing in the way of implementing the majority of superconducting applications. On the other hand, superconductivity has seldom failed to deliver when called upon to perform tough technological tasks. Many remaining problems in superconducting technology only await the application of resources and talent to find their solutions.

The economic feasibility issue constitutes a different story. We are not primarily talking about the monetary costs of liquid helium itself, even though liquid helium costs 10 times as much as liquid nitrogen. We are talking about the costs incurred by the complexity and inconvenience that automatically accompanies helium cooling, an issue we explore in detail later on. Nevertheless, as we have seen, there are jobs that only superconductors can do and there are customers who are already willing to pay the price of superconductivity. The costs associated with liquid-helium cooling do not deter medical diagnosticians, high-energy physicists, radioastronomers, satellite designers, and a host of other eager users of advanced superconducting technology. On the other hand, liquid helium dramatically cools the enthusiasm of transportation engineers, electrical power experts, and many others for whom superconductivity represents a tantalizing but impractical prospect for innovation.

The year 1987 saw the discovery of superconductivity above liquid-nitrogen temperature and with that discovery, the superconductivity business will never be the same. It may be quite some time before high-temperature superconductors actually enter the marketplace in the form of advanced electrical and electronic products, but they have already entered the imagination of superconducting technologists around the world. Superconducting devices, machines, and systems can now be conceived with the promise of operation without the costly constraints of liquid-helium cooling. The economic promise of such future superconducting applications has given the superconductivity industry a boost that will propel it into the twenty-first century.

CHAPTER 13

The Breakthrough

For a few weeks in February, 1987, hundreds of previously staid scientists behaved like espionage agents from a B-movie. Late-night phone calls and furtive meetings centered around "the secret formula"; everybody wanted it and only a few people had it. All the cloak-and-dagger activity surrounded the biggest single event in the superconductivity world since Kamerlingh Onnes first dipped mercury wires into liquid helium. Superconductivity had broken the liquid nitrogen temperature barrier.

For diehard scientists, crossing this threshold was like running the four-minute mile, breaking the sound barrier, and landing on the moon, all rolled into one. Perhaps what made the superconducting breakthrough such big news was that most scientists thought that it would never happen. But the course of research that led up to the breakthrough in high-temperature superconductivity was pursued by people who did believe it was possible.

Earlier, we followed the trail of discovery of new superconductors and saw how it was blazed by scientists like Bernd Matthias. By 1971, following Matthias' rules, researchers had produced a 23 K superconductor made of an intermetallic compound material. For many more years, others tried to follow that same road to even higher temperature superconductors, but with no success. The path that led to success had its origins in the 1960s, when a few scientists began to find superconductivity in entirely new places.

The Road to the New Superconductors

In 1964, a research group headed by James Schooley at the National Bureau of Standards in Washington reported superconductivity in a crystalline compound called strontium titanate. They were testing a prediction by a theorist named Marvin Cohen. Although the superconductivity that they found occurred all the way down at 0.7 K, it was nonetheless a remarkable discovery because strontium titanate is an oxide—a chemical combination of oxygen with other elements. Ordinary rust is an oxide of iron. We would not expect oxide materials to be superconductors, because they are generally electrical insulators—including such familiar examples as crystalline quartz and sapphire (which are silicon and aluminum oxides, respectively). Yet here was a transparent crystalline material—which, at the time, was even used to make synthetic diamond jewelry—that was able to superconduct at low temperatures. Strontium titanate has no particular value as a superconductor, on account of its low transition temperature, but its discovery initiated interest in a whole new class of materials as candidates for superconductivity.

David Johnston, a student of Bernd Matthias at the University of California, San Diego, discovered a related oxide superconductor with a truly respectable transition temperature in 1973. The material—lithium titanate—superconducted at 13 K. With this discovery, Johnston dispelled a major misconception about oxide superconductors, namely that their superconductivity was limited to very low temperatures. Thus, 2 years later, when Arthur Sleight at DuPont discovered another superconducting oxide with a transition temperature of nearly 14 K, no one was terribly surprised. The electrical characteristics of thin films of that material, barium-lead-bismuth-oxide, turn out to be quite sensitive to radiation. For that reason, the material has proven to be useful for building infrared detection devices and a number of scientists have continued to study it ever since.

Despite these discoveries, the oxide superconductors were never considered to be the prime candidates for higher temperature superconductivity. That honor passed from the intermetallic compounds—formed only from metals—to materials called organic conductors, then to unusual thin-film materials, and finally back to intermetallics. The frustrating failure to increase critical temperatures beyond 23 K in any kind of material eventually prompted physicists to speculate that higher temperatures were simply incompatible with superconductivity. Less pessimistic theorists suggested that higher temperature superconductors could exist, but only if the superconductivity resulted from some entirely new mechanism different from the BCS pairing phenomenon. For this reason, exotic materials like organic conductors were thought to be ideal candidates for higher temperature superconductivity.

Over the years, researchers found a variety of unexpected materials to be superconductors. Robert Hein found superconductivity in a semiconductor—an alloy of germanium and tellurium—back in 1964. Other researchers reported superconductivity in a material composed of long chainlike molecules of sulfur and nitrogen in 1975. By 1980, even organic conductors were shown to be capable of superconductivity. Organic conductors are large carbon-based molecules that more closely resemble biological matter than metal. Such materials form in chainlike structures that can effectively conduct electricity in only one favorable direction along the chain. Some of these materials actually contain no metallic atoms at all. Nonetheless, they conduct electricity like metals and a few of them actually superconduct, with the highest known transition temperature at 8 K.

Since the late 1960s, scientists have reported breakthroughs in high-temperature superconductivity on a number of occasions. A Russian report attracted a great deal of attention in 1978, claiming superconductivity in a common chemical—copper chloride—at temperatures close to an astounding 140 K. At the time, Bernd Matthias branded such reports as deliberate attempts at deception, although a young protegee of his—Paul Chu—cautioned that some reports of high-temperature superconductivity might prove to be valid. Over the next few years, many scientists continued to search for the superconductivity in copper chloride, but no one ever succeeded in obtaining any reproducible and incontrovertible results. In the ensuing years, other materials continued to surface as alleged high-temperature superconductors, but again and again, the results could never be authenticated.

Consequently, scientists who claimed to find high-temperature superconductivity gradually came to be regarded in the same vein as "the little boy who cried wolf." Nobody was inclined to believe their results and the scientific community gradually lost interest in the search for new superconductors. It was against this backdrop that a pair of researchers at IBM's Zurich laboratory embarked on a course of research that would ultimately revolutionize the superconductivity world.

The Zurich Discovery

Karl Alex Mueller, head of the Zurich lab's physics department and a professor at the University of Zurich, had worked on a variety of problems in the electronic properties of materials since joining IBM in 1963. By the time Johannes Georg Bednorz received his doctorate in 1982 and joined the IBM research staff on a full-time basis, Mueller had already cultivated an interest in superconductivity. He became convinced that researchers searching for higher temperature superconductors were on the wrong track. During a discussion at a conference in Italy with Harry Thomas, his former IBM supervisor, Mueller became persuaded that the metal oxides were the right place to look.

Upon his return to Zurich, Mueller enlisted the aid of Bednorz to help him undertake a systematic search of the metallic oxides for superconductivity. The 14 K superconductivity in barium-lead-bismuth-oxide demonstrated the genuine promise of these materials. Mueller and Bednorz reasoned that some kind of unusual electronic behavior on the part of an oxide's atoms would be needed to free up the normally bound electrons of an insulator for the conduction of electricity. They focused on metals with so-called mixed-valence states. These were metals whose atoms could change their chemical behavior depending on where they were found in a compound. The most promising candidates of this type were metal oxide compounds of copper and nickel.

Mueller and Bednorz experimented with a number of obscure materials—compounds like lanthanum nickel oxide—by trying to alter the properties of such substances by adding small amounts of other elements. The work went on for 2 years as a sideline project; IBM management, although supportive of independent research, was only vaguely aware of what the two scientists were up to. Management knew that Mueller and Bednorz were studying metal oxides, but the search for new superconductors was cloaked in secrecy. Working at odd hours and in between tasks for his primary projects, Bednorz managed to produce and test scores of materials, looking for superconductivity. After 2 years, the lack of progress left him very frustrated.

When research seems to be leading nowhere, scientists often head for the library for new inspirations. At the end of 1985, Bednorz decided to scan the literature for ideas about copper compounds; nickel was looking like a dead end for superconductivity. A recent report from the University of Caen in France contained some exciting news. Claude Michel, L. Er-Rakho, and Bernard Raveau had investigated a compound of lanthanum, barium, copper, and oxygen that seemed to have some interesting properties. The French chemists were not looking into superconductivity, but they had successfully synthesized a rather unusual copper oxide material.

Studying the French results, Mueller and Bednorz realized that this compound was just the sort of mixed-valence metal oxide material they were most interested in. Chemists make these oxide compounds using a straightforward technique know as solid state reaction. They carefully mix together chemical powders containing the desired elements and heat the mixture to very high temperatures in order to promote chemical reactions. The resultant solid mass then gets reground, pressed into pellet form, and re-baked to solidify it for use in experiments. Making such samples is tedious and time consuming, but Bednorz produced sample after sample in this way. He began synthesizing samples of the French compound, varying the amount of barium it contained.

On the evening of January 27, 1986, he attached leads to a blackish pellet of a version of the French material and immersed it into a liquid helium dewar. The resistance of the material dropped sharply at about 11 K. Bednorz went home that night thinking he had succeeded in his long search, but he wondered if anyone else would be convinced. The next day, he repeated the measurement and obtained the same result. For the next couple of months, Mueller and Bednorz worked on improving the material. Eventually, they had samples with resistances that dropped above 30 K. The resistance didn't drop all the way to zero in these samples; they were of rather poor quality. Nonetheless, Mueller and Bednorz knew they were on to something big.

The climate of distrust surrounding reports of high-temperature superconductivity made Mueller and Bednorz very cautious about revealing their discovery. The resistance changes would not be enough to convince all the skeptics; only the addition of a Meissner effect measurement could provide the necessary proof of superconductivity. Thus they decided to await the arrival of a SQUID suscep-

Nobel Prize winners J. Georg Bednorz and K. Alex Mueller. (Courtesy IBM)

tometer they had ordered, a sophisticated instrument for making magnetic measurements. To establish their precedence, however, they submitted a cautious article to the German journal *Zeitschrift fur Physik* on April 17. The paper, entitled "Possible High T_c Superconductivity in the Ba-La-Cu-O System," appeared in print in September, 1986. A month before, the susceptometer had arrived at Zurich and the new superconductors had passed the Meissner test. Mueller and Bednorz were confident of their findings.

The announcement that the 23 K barrier had been broken did not initially arouse much interest. There had been too many false alarms before and the original paper was not convincing enough. However, the news gradually began to spread. A visiting scientist from the University of Tokyo, Masake Takashige, helped Mueller and Bednorz make some of the early magnetic measurements in October. By the second week of November, the superconductivity group in Tokyo was able to reproduce the Zurich results. For the first time, a super-conductor above 30 K had passed the critical scientific test of re-producibility.

In America, the first research group to start working with the new superconductor was that of Ching-Wu Paul Chu at the University of Houston. Chu's main interest for many years had been novel superconducting materials; he had been involved in the investigations of copper chloride in the early 1970s while at Cleveland State University. Chu learned many of his methods from Bernd Matthias, his graduate school mentor at UC San Diego, and thus he subscribed to Matthias' policy of always paying close attention to the literature. A routine literature search by one of Chu's students yielded the Mueller–Bednorz article. As soon as he saw the article, Chu became excited by the Zurich results and told his group to drop everything else and study the new material. The Houston group produced their own sample of the Zurich compound on November 12, one day before Shoji Tanaka's group in Tokyo. The worldwide effort in high-temperature superconductivity had begun.

Despite the publication of the Zurich results and the work in Houston and Tokyo, most scientists were still unaware of the break-through at the end of November, 1986. Finally, on December 4, the situation changed dramatically. On that Thursday, the annual winter meeting of the Materials Research Society was going on in Boston. One of the sessions featured several papers on the older oxide super-conductor—barium-lead-bismuth-oxide—including reports from Paul Chu and from Koichi Kitagawa from the Tokyo group. At the end of his presentation, Chu announced his recent results confirming the Zurich discovery. Kitagawa then stood up and showed the Tokyo results as well. The word was out; there really was a new high-temperature superconductor.

From that point on, the number of research groups working with the new compound mushroomed. Major laboratories in the United States, in Japan, and in China all began synthesizing the Mueller-Bednorz compound and trying to improve upon it. Several groups quickly identified the precise chemical composition actually responsi-ble for the superconductivity; the original Zurich samples in fact con-tained a mixture of several different oxide compounds. The part that was a 35 K superconductor turned out to be one particular compound made of lanthanum metal, copper, oxygen, and a little bit of barium. The barium acts as a dopant—a material present in small quantities to

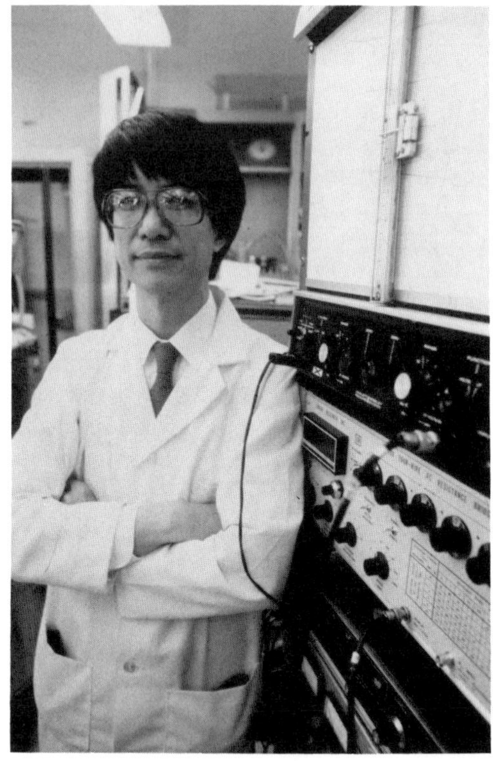

C. W. Paul Chu. (Courtesy University of Houston)

influence the behavior of the compound. Chemists write the super-conductor's now-famous formula as $La_{1.85}Ba_{0.15}Cu_1O_4$. It has become famous because IBM splashes the formula across full-page ads in newspapers and magazines, hailing the achievement of their researchers. But in December, 1986, scientists struggled to identify the formula and determine the crystal structure of the compound. The structure of the new superconductor was a close relative of that of the first oxide superconductor: strontium titanate. Chemical structure experts call these compounds *perovskites*, which are crystal structures that occur very commonly among natural minerals.

During that month, apart from the precise identification of the Zurich superconductor, a number of research groups experimented with adding different amounts of barium to the Mueller and Bednorz material, as well as with substituting other elements in the compound. With a substance that contains four different elements, the possibilities for variations are nearly endless. A particularly productive line of research originated at Chu's group in Houston. Chu had studied the effects of high pressure on materials while he was a postdoctoral researcher at Bell Laboratories. To continue such work, he had established the capability to do high-pressure measurements in his Houston lab. Over the years, a number of materials have been found to superconduct only when under extreme pressure, while some other superconductors simply have higher transition temperatures under pressure. Intense pressure changes the spacing between the atoms in a material—it squeezes the crystal lattice together a bit—and, occasionally, this effect serves to enhance the interactions that cause superconductivity. Chu thought it worthwhile to try this trick with the Zurich superconductor. Remarkably, not only did intense pressure work, it worked better than it had on any other superconductor. Generally, pressure effects bring about only small changes in superconducting transition temperatures. In the new superconductor, the Houston team was almost immediately able to drive up the onset of the superconducting transition (the temperature at which the resistance begins to drop) to over 40 K; and by the end of December, they reported signs of superconductivity at temperatures as high as 52 K when they applied pressures of nearly 200,000 pounds per square inch to the material. The world record for superconductivity was being broken at a dizzying pace.

Buoyed by these successes, Chu began to think about chemical ways to simulate the effects of high pressures in order to create a higher temperature superconductor that did not require the immense pressure. The goal was to get the atoms in the compound closer together; Chu decided to replace the barium atoms with smaller strontium atoms to achieve the result. Several other labs pursued the same line of reasoning and by late December, groups at Bell Laboratories and at Tokyo reported superconductivity at 40 K in the lanthanum-strontium-copper-oxide material with no pressure required. At this point, the Bell group produced a sample with no detectable resistance at 36 K, the highest value ever reported. Some groups made the compound with calcium instead of strontium, since calcium atoms are even smaller than strontium atoms. Unfortunately, this only reduced the superconducting temperature. Putting pressure on the strontium compound also didn't raise the superconducting temperature. Apparently, the researchers had gotten all they could out of manipulating atomic sizes.

Breaking the Liquid-Nitrogen Barrier

By the beginning of January, 1987, the efforts seemed to change direction. At that point, hundreds of scientists were in a frantic race to produce better superconductors. They were changing the proportions of the elements in the new compounds and changing the elements themselves. Soon there were more samples than could be measured; little plastic boxes containing the pellets were piling up in many laboratories. In Houston, several undergraduate students were recruited to help mix and grind powders as Chu's group tried to work its way through a long list of element substitutions. The list was so long that it had been divided up between the group in Houston and another team at the University of Alabama in Huntsville, headed up by Maw-Kuen Wu, a former student of Chu.

Substituting one element for another in these compounds was not a random activity; the elements chosen had to share common features, particularly their chemical valence—the number of electrons in an atom that participate in chemical bonding. Thus, substitutions for barium could only be made from among elements with the same valence such as strontium and calcium. Similarly, lanthanum could only be replaced by elements from a group called the rare earths. One of these rare earths—the metal yttrium—was on the list of compounds that Wu's group was investigating. By the end of January, it was yttrium's turn in the Alabama team's series of experiments.

When the first pellet of yttrium-barium-copper-oxide came out of the oven, it was a greenish uneven-looking material. The lanthanum superconductors looked more like charcoal, so this new yttrium material didn't seem too promising. Chemical compounds get their color from the interactions of light with the electron energy bands we discussed earlier. Conductors generally don't absorb light in such a way as to produce visible colors. Nevertheless, the Alabama group checked out the latest sample.

Wu's graduate student, James Ashburn, cooled the green pellet on January 28. Incredibly, its resistance disappeared at 93 K. Wu recalls shaking from excitement as he watched the new substance became fully superconducting in a bath of 77 K liquid nitrogen. The Houston-Alabama team had hit the jackpot in superconductivity.

After frantic phone calls to Houston and to the airlines, Wu and Ashburn packed up the precious sample and took the first plane to Houston. During the next few days, Chu and Wu reproduced the results and even raised the temperature by a few degrees. Superconductivity above the temperature of liquid nitrogen had been something that physicists only dreamed about; now it was a reality. Chu and his team contemplated their next move.

They wanted to find out as much as possible about their discovery before revealing it; when larger, better equipped laboratories found out about the new material, they would surely take over the bulk of the research. So although they had the advantage of exclusivity, the Houston group wanted to identify the actual superconducting compound as well its chemical structure. The University of Houston did not have all the facilities that Chu needed. For certain measurements, he would have to send out the samples. In the meantime, he wrote an article describing his results. The problem was how to submit the article for publication without immediately revealing what he had done.

Chu wanted to send his article to *Physical Review Letters*, the most prestigious journal for the rapid reporting of significant results in physics. A scholarly article must always be reviewed by expert referees to judge whether it reports competent and worthwhile results. Therefore, Chu struck a deal with the journal in which he would personally select the two referees for the paper, people he could trust with his secret. Despite a standard code of confidentiality for reviewers, Chu was still wary. Nevertheless, on February 6, he sent copies of the article both to the journal and to the two referees. During the following week, despite the precautions, the rumors began to fly. By the end of that week, scientists around the world were talking about the greenish 94 K superconductor that was made in Houston.

Back on January 12, Chu had already filed a patent disclosure for a new high-temperature superconductor. He had seen clear evidence of 90 K superconductivity in magnetic measurements on lanthanum-barium-copper-oxide pellets (as we will see, a whole family of 90 K superconductors exists besides the original yttrium compound). Finding zero resistance in the yttrium compound 2 weeks later was the final proof of Chu's claim. The nature of what can be legally protected in a discovery of this kind constitutes a thorny legal question. The precise claim in the patent will therefore probably become the center of an intense legal battle in the future. Two weeks after the Alabama breakthrough, Chu was ready to face the world. The official announcement of the discovery came on February 16, 1987, at a press conference in Houston. Here Paul Chu announced that his team had synthesized the world's first above-liquid-nitrogen-temperature superconductor and that the composition would be revealed in his article to be published at the end of the month. He would keep the formula secret until that time.

The guessing game began for the identity of the secret formula. The original manuscript of Chu's article sent to the journal contained a crucial (typographical?) error. Instead of containing the letter "Y" symbolizing the element yttrium, in every instance the formula was written with "Yb," the symbol of another rare earth element, ytterbium. Despite the confidentiality code, the ytterbium-barium-copper-oxide rumor proliferated in the scientific community. As a result, ytterbium sales to laboratories skyrocketed.

On February 16, the *Houston Chronicle* published an article about the exciting superconductivity discoveries made at the local university. The reporter who wrote the article had interviewed a university dean who had seen Chu's patent disclosure. In the course of the interview, the dean told the reporter about the remarkable new yttrium-barium-copper-oxide compound being made by Chu. Consequently, the correct formula appeared in the article. Astonishingly, no one in the United States appeared to notice. For the next 10 days, scientists continued to investigate ytterbium compounds and otherwise speculate about the true identity of Chu's compound.

Foreign scientists did appear to get the word, however. By February 21, research groups in China and Japan had begun to make Chu's superconductor. In fact, the correct composition was published in the *People's Daily* in China on February 25. This time people took notice, but by then the official word was coming out everywhere. On February 26, Chu related the details of his work at a standing-room-only lecture at the University of California at Santa Barbara. He also air-expressed copies of his article to ten major laboratories around the country. Word of the formula spread like wildfire. That weekend saw hundreds of laboratories frenetically preparing samples of the correct new superconductor. Within days, Chu's results were reproduced a hundred times over.

In the flurry of activity that commenced late in February, it became increasingly difficult to determine just who did what first. The Alabama-Houston team clearly were first to see superconductivity above 77 K but, ironically, they were not the first people to actually synthesize the high-temperature superconductor.

Questions of precedence come up with the Bednorz-Mueller discovery as well. A Russian paper in 1978 reported on the properties of some rare earth oxide compounds including one nearly identical to the lanthanum-strontium-copper-oxide 40 K superconductor. The Russian researchers never looked for superconductivity in their samples, but if they had, they probably would have found it. Similarly, the French group lead by Bernard Raveau—whose work inspired the Zurich discovery—had also synthesized compounds in the early 1980s that turned out to contain high-temperature superconductors. Why don't these scientists receive the credit for the discovery? Simply because they were not looking for superconductivity and didn't find it. Kamerlingh Onnes did not discover mercury; he discovered superconductivity in mercury. Similarly, only Bednorz and Mueller discovered superconductivity in lanthanum-barium-copper-oxide.

The stories of earlier samples of Chu's compound are even more frustrating for those who came so close. In the frenzy of sample preparation at the end of 1986, scientists had made quite a few samples that they had never managed to measure. Some of these samples turned out to be the same as Chu's superconductor. One such sample at a government lab was even briefly measured on January 6, 1987, but the results were discarded as being preposterous. Another such early sample did manage to make a little history of its own.

On January 3, 1987, Jean-Marie Tarascon, a physicist at Bell Communications Research, produced several pellets of the yttrium-barium-copper-oxide compound. In the rush of activities at the time, the samples were never measured. The blotchy green pellets were clearly poor quality material and did not warrant the time required for the measurements. On February 26, the Bellcore group received their copy of Chu's paper. That evening, they tested Tarascon's eight-week-old samples and found that two of them superconducted above 90 K. The next morning they wrote up the results of the previous night's measurements and late that afternoon, drove off to the offices of *Physical Review Letters* to submit their manuscript. Finding the office closed, they persuaded a security guard to acknowledge receipt of the article on that date. The resultant publication was the second article to appear about the new superconductor and the first to publish independent verification of Chu's work. This article established a new precedent for speed, as it reported the results of scientific research that had not even been conceived the day before the article was submitted for publication.

There are probably other such stories of missed opportunities.

Most of them will remain known only to their embarrassed principals. In any event, the important distinction remains that Chu's team was in fact the first to demonstrate superconductivity in the compound. History will remember that accomplishment long after the missed-opportunity anecdotes are forgotten.

The last week of February saw many research groups working hard at identifying the new superconducting compound. Just like the original Zurich material, Chu's compound was actually a mixture of different oxides—a mixed-phase material, as the chemists say. Material scientists soon determined that the material was mostly made up of two constituents: a green compound and a black compound. The question was: which one was the superconductor?

Chu finally sent out some samples to be analyzed on February 20. They went to the Geophysical Laboratory in Washington, D. C., an institution particularly well-equipped to analyze the properties of the rocklike samples. In chemistry terms, they needed to isolate the phases; what was the green stuff and what was the black stuff? During the next two weeks, material scientists at the Geophysical Laboratory, at Bell Labs, at IBM, at Argonne National Laboratory, and elsewhere all labored to identify the new compound. Scientists employed sophisticated techniques that could probe microscopic grains from within the pellets. By the end of the first week, they started getting answers.

Chemists figure out the contents of a new substance with a variety of analytical techniques, but they almost always use X-ray diffraction to determine how the atoms go together. X-ray diffraction works by shooting an X-ray beam into a material and watching how the X rays come back out. X rays have such small wavelengths that the planes of atoms in a crystal very effectively deflect an X ray as it travels through the crystal. By carefully monitoring the direction and the strength of the X rays that emerge from a crystal, scientists can infer the precise arrangement of atoms that deflected the X rays. When a sample contains a number of different kinds of crystals, the X rays scatter off in all sorts of directions, making the analysis process extremely complicated. This was the sort of problem that the researchers had with Chu's new superconductor.

Despite the difficulties, several labs figured out the composition and structure of both the green and black compounds in the superconducting samples. As it turned out, the green stuff was not the superconductor; in fact, it was an insulator. Workers at the Geophysical Laboratory figured out its complex crystal structure in a remarkably short amount of time only to find out not only that it was not the superconductor, but that the French group of Michel and Raveau had already analyzed the compound several years before.

Thus scientists realized that the black compound was Chu's new superconductor. By the end of February, all the major research labs had determined that the superconducting compound has the formula $Y_1Ba_2Cu_3O_7$, which immediately prompted scientists to dub the material "1-2-3". The Geophysical Laboratory probably was first to find this out, but the timing was close. The crystal structure of the new compound generated a great deal of interest. Once again, the material was part of the perovskite family, just like the Zurich superconductor and strontium titanate. In Chu's material, the structure was even more complicated with many more distinct layers forming the compound. We explore the details of the structure later on.

The Woodstock of Physics

The situation at the end of February, 1987, was remarkable. Hundreds of scientists from several different fields—perhaps even a couple of thousand scientists—were all studying the same material. As a result, there was a virtual information explosion. The labs that identified the actual superconductor immediately shifted their efforts to making samples of just the black compound. Researchers set up cots in laboratories and brought in meals as they produced and analyzed new samples on an around-the-clock basis. Events were taking place far too rapidly to be reported in scientific journals; the latest results appeared in the *New York Times* and elsewhere in the popular press. Quite often, the exact same experiment would be performed in 50 different laboratories within a period of hours. Researchers made new discoveries almost daily.

The first real chance for scientists to congregate and share their findings came in the week of March 16 when the American Physical Society held its annual week-long solid-state physics meeting in New York. The rules of the conference require that brief descriptions of work to be presented be submitted 4 months before the meeting. As a result, only one 10-minute paper reporting early measurements on the Zurich superconductor was scheduled. Meeting organizers realized that they had to organize a special session in order to accommodate the flood of last-minute results on the new materials. They scheduled the session for Wednesday night, March 18.

Attenders of the conference were asked to turn in a special form indicating their interest in the special session. The organizers quickly realized that more people wanted to attend than the meeting room could hold. The session was scheduled to begin at 7:30 p.m. and the doors would open at 6:00. At 4:00, the crowd began to form outside the meeting room. During the next hour, the crowd of scientists filled the lower lobby and began to pile up on the escalator leading to the main lobby in the New York Hilton. By 6:00, it was a scene out of a stadium rock concert; over 3,000 people pressed toward the meeting room doors.

When the doors opened, a human tide swept through. There were at least two people for every chair in the room. Within minutes, every seat was taken and hundreds of people were lined up against the walls and in the aisles. The air-conditioning system could not cope with the huge crowd and everyone sweltered in the packed hall. Conference organizers had to coerce large numbers of scientists to leave the room in order to comply with fire regulations. Even a Nobel Laureate was evicted. The hotel set up television monitors in several lobbies and arranged additional seating for the enormous turnaway crowd that spilled into all the adjacent spaces. Newspapers later called the event "the Woodstock of physics."

During the actual session, scientists from around the world reported an extraordinary amount of research—all of it a few weeks old or less. The biggest names in the field—Alex Mueller, Paul Chu, Shoji Tanaka from Tokyo, Zhao Zhongxian from Beijing, and Bertram Batlogg representing Bell Laboratories—made the first presentations. The huge crowd gave a tremendous ovation to these scientists whom the session chairman, Neil Ashcroft, introduced as those "who set this magnificent engine running."

Throughout the evening, dozens of scientists reported the results of their studies of the new superconductors. At least eight additional compounds had been found to superconduct above liquid nitrogen temperatures. Two groups—from IBM and Stanford University— even reported making superconducting thin films of the new materials. The presentations went on until 3:15 a.m., and small groups of scientists continued their discussions in the lobbies until dawn. It was a night to remember.

If a mad rush had already been in progress before the New York

The scene at the all-night session of the American Physical Society meeting.
(Courtesy David Kalson, AIP)

meeting, sheer pandemonium broke out in its aftermath. Thousands of scientists returned to their home laboratories, eager to work on the new compounds. The list of things to do was nearly endless. People wanted to know what caused the superconductivity at such high temperatures, they wanted to measure all the superconducting properties of the new compounds, they wanted to try to make better samples of the materials than the crumbly pellets they had already seen, and they wanted to find even higher temperature superconductors.

Late January has become the breeding season for high-temperature superconductors and, right on schedule, January, 1988, ended with scientists at the University of Tsukuba in Japan reporting superconductivity in yet another kind of copper-oxide compound. This time the material contained four different metallic elements including bismuth. Their initial results indicated a transition temperature above 100 K. At first, people assumed that another premature report had made the press, but Paul Chu's group at Houston quickly reproduced these results and produced samples that showed signs of superconductivity at 115 K. In a state of déjà vu, the research community embarked on an effort to identify and characterize the new material.

Within weeks, labs around the world were busily studying this new high-temperature superconductor.

Soon thereafter, researchers at the University of Arkansas announced the discovery of still another copper-oxide superconductor, this time containing the element thallium. Within a few weeks, the verified record for a superconducting transition temperature climbed to 125 K, thirty degrees higher than Chu's compound. Only time will tell whether this compound, the bismuth compound, Chu's original compound, or perhaps an entirely new material will prove to be the most important high-temperature superconductor for future applications.

Starting in early 1987, there has been a string of reports of superconductivity at considerably higher temperatures than either the 90 to 94 K universally seen in Chu's original material or the 110 to 120 K measured in the bismuth and thallium compounds. Chu himself reported tantalizing hints of superconductivity—abrupt decreases in sample resistances—at 140 K, at 160 K, and even at 240 K. During the summer of 1987 many research groups reported promising drops in the resistance of samples at temperatures above 200 K. Scientists went directly to the press with these results, usually bypassing the scientific journals. The enthusiastic coverage resulted in a general public impression that superconductivity had been pushed up to temperatures close to room temperature (290 K).

Despite the many reports of such higher temperature superconductors, none were able to be reproduced. Were the scientists who claimed to find new higher temperature superconductors all frauds? Certainly not. At this point at least, reports of higher temperature superconductivity are rather like sightings of UFOs. Researchers did see *something* in their laboratories, but probably not what they thought they did. How could scientists mistake something else for superconductivity? We can point to several common mistakes.

Scientists measure resistance by attaching four wires to a sample, passing current through two of the wires, and measuring voltage across the other two. In practice, some materials are difficult to attach wires to because of their rough, uneven surfaces. Pellets of oxide superconductors are a prime example. Individual wires coming loose from a sample during measurements can produce sudden voltage drops that mimic the effects of superconductivity. Quite a few scientists have thought they had a new superconductor because of this kind of problem. Another kind of false alarm for superconductivity can come from materials whose crystal structures change abruptly at a certain temperature. Such structural phase transitions can be accompanied by dramatic changes in resistance, which can look just like a superconducting transition, if one is not careful. The difference is that the resistance does not go to zero in these materials but just to a much lower value than before.

There are many other possible explanations for false reports of superconductivity. The way to avoid making such reports is to perform careful and repeated tests to verify the experimental results. At a time when the spotlight was on superconductivity, scientists were not taking the time to perform the necessary tests. Also, many of the scientists working on the new materials had no previous background in superconducting measurements and were prone to errors of inexperience. Any material that really turns out to be a new superconductor must be able to pass a number of critical tests whose results can be reproduced by other scientists. The wave of new discoveries in early 1988 can only spur on the worldwide efforts of scientists trying to find even higher temperature superconductors.

Aside from the efforts to find new and still better superconductors, growing numbers of scientists have embarked on a course of basic scientific research on the new materials. Moreover, superconductivity at liquid- nitrogen temperature holds such tremendous promise for applications that it has prompted an even larger effort in applied scientific research. Next, we discuss the nature of the new high-temperature superconductors and the problems that scientists need to solve in order to use them.

CHAPTER 14

The New Superconductors

The discovery of high-temperature superconductivity has been hailed as one of the most important scientific events of the century. The amount of frenzied activity and excitement generated by the discovery supports this claim. Yet high-temperature superconductivity might be thought of as an enabling discovery rather than a ground-breaking discovery since superconductivity itself was already well known and its potential widely investigated before the latest break-throughs. The immediate excitement surrounding the new supercon-ductors stems from their potential for enabling old ideas to be brought to fruition. However, as we will see, the new superconduc-tors may also enable a new revolution in scientific imagination.

During the first year of high-temperature superconductivity (1987), numerous questions have been formulated about the subject and some of them have even been answered. The most general ques-tions that come to mind are: what are these new superconductors, how do they work, what needs to be done to use them, when will they start being used, and what will we do with them? In the next two chapters, we try to work our way through this ambitious list of inquir-ies. In some cases, the answers are unclear; in others the answers are obvious.

We first consider the nature of the new superconductors. *High-temperature superconductor* is a fickle term. It used to mean any super-conductor with a transition temperature above 10 K. Now it probably means anything with a transition temperature above 40 K or maybe even 90 K. Perhaps this compares to the fact that any professional basketball player above 6 feet-8 inches in height used to be considered a "big man," whereas now such players are of average height. In any case, 1987 saw the discovery of superconductivity well above the temperature of liquid nitrogen—77 K.

All about 1-2-3

The high-temperature superconductors are not metals or inter-metallic compounds at all; they are oxides. According to crystal chemists, the compounds are classified as ceramics. Aluminium oxide, quartz, and even ordinary table salt are examples of ceramic materials, so defined by virtue of the kind of bonding that holds their constituent atoms together. For most of us, the term *ceramic* makes us think of pottery that is fired in a kiln, and in fact, the new high-temperature superconductors were first made in just this way. Ceramic oxides are a far cry from the simple metals that were the proving ground for superconducting science and technology. The new oxide superconductors challenge a theoretical understanding of superconductivity that took many decades to develop.

At least two different groups of copper oxide compounds exhibit superconductivity above 90 K. The first compound discovered by Paul Chu bears little resemblance to the superconducting metals that Bernd Matthias once studied. Such high-temperature superconductors have a fascinating crystal chemistry.

The crystal lattice of a solid can be thought of like a brick wall; the structure consists of repeated building blocks known as the *unit cell*. The unit cell of a solid, like a molecule of a gas, contains a certain set of atoms that make up a compound. In a unit cell, the atoms are arranged in a certain way so that repeating the cell over and over in three dimensions produces the compound's crystal structure. Forming a complete crystal by replicating the unit cell is analogous to stacking bricks to form a wall. To cite a simple example, the unit cell for sodium chloride—ordinary table salt—consists of a sodium atom and a chloride atom sitting in specific positions. Replicating that cell produces the salt crystal.

The new superconductors have a far more complicated unit cell. The cell contains one atom of a rare earth metal, two barium atoms, three copper atoms, and seven oxygen atoms. (We talk about rare earth metals momentarily.) The number of atoms of each metal

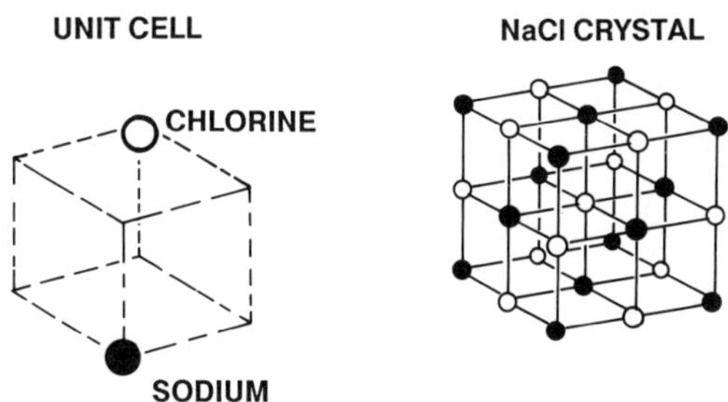

Salt crystal showing unit cell and complete crystal.

element in the superconductor gives these compounds the name "1-2-3," which has become the familiar way to refer to any of them.

Most familiar superconductors have crystal structures with cubic symmetry. When you rotate a cube 90 degrees in any direction, you we see another identical face. Similarly, rotating a unit cell with cubic symmetry 90 degrees in any direction does not change its appearance. The former high-temperature champions—the niobium-based A15 compounds—indeed have cubic symmetry, as the diagram on page 115 shows. But the new superconductors are not cubic at all. The unit cell has a rectangular shape but no two dimensions are the same size. Crystal chemists describe the structure as an *oxygen-depleted layered perovskite crystal*. This imposing mouthful can be best understood with the aid of a diagram.

The 1-2-3 superconductors form as layers of copper and oxygen atoms sandwiched between layers containing the other elements in the compound. Some of the copper–oxygen layers are made of planes of atoms while others are made of chains of alternating copper and oxygen atoms. Some scientists believe that the copper–oxygen chains are crucial to superconductivity and others think that the planes are the important structures. In either case, only the layers containing oxygen and copper atoms appear to be important for the electrical properties of 1-2-3. Nevertheless, there are three metal elements in the compound. Although scientists believe that the metals themselves don't play a direct role in the electrical conductivity of 1-2-3, they are not unimportant. The structure that atoms take on to form a crystal comes about from a delicate balance of electrical forces between them. The spacing and placement of the atoms in the crystal

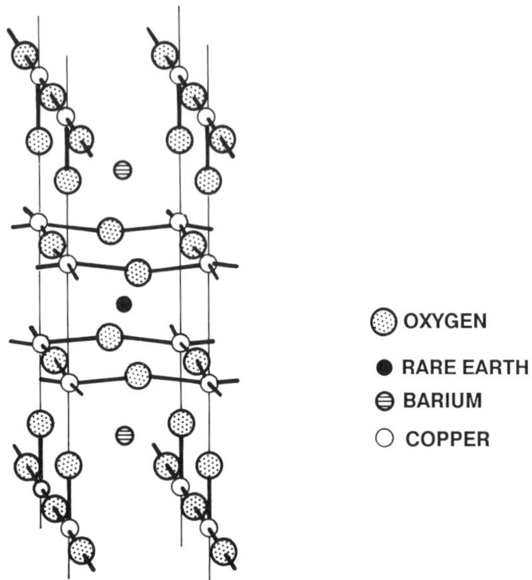

The unit cell of yttrium-barium-copper-oxide (1-2-3).

depends on atomic sizes and numbers of electrons. Crystal formation has its own elaborate accounting practices.

In the 1-2-3 compound, the first element that scientists list in the chemical formula—the rare earth metal—turns out to be the least critical one. Since Chu's discovery of 1-2-3 containing yttrium, scientists have synthesized additional 1-2-3 compounds substituting for yttrium with other rare earth elements such as gadolinium, holmium, europium, erbium, and a host of others. The compounds have the remarkable property that their superconducting behavior does not depend on which rare earth element it contains. Rare earth metals are not so rare at all, it turns out, but they are unfamiliar to most of us. All are physically and chemically quite similar to one another and are therefore quite difficult to separate. They tend to be soft silvery-white metals that oxidize readily in ordinary air. Most were discovered during the nineteenth century and virtually all are found in monazite, the same mineral Kamerlingh Onnes used for a source of helium gas. It somehow seems fitting—although completely coincidental—that the mineral that played a crucial role in the original discovery of superconductivity should also be a prime source of a component of the new superconductors. Until the middle of the twentieth century, it was most common to use the rare earth metals in combination, since that is how they are found in nature. One such mixture has the name *misch metal*, a material used for cigarette lighter flints on account of the pyrophoric property of the rare earths—a tendency to burst into flame when fresh surfaces are exposed to air. Only more recently has it become practical to separate the rare earths from one another and use them individually.

As far as we can determine, the rare earth atom in 1-2-3 more-or-less acts like a spacer between copper–oxygen planes; its chemical properties don't figure into the electrical properties of the superconductor. Thus, there exists an entire family of 1-2-3 compounds that all superconduct above 90 K and differ only in the identity of the rare earth element.

The 1-2-3 compound is a complicated material. It contains four different elements that are arranged in a very particular fashion. There was certainly no reason to expect superconductivity from such a substance. More pointedly, there was no reason to expect conductivity at all from the 1-2-3 material. It is an oxide like quartz or sapphire, not a metallic compound, and most oxides are insulators. Yet, in the normal state, 1-2-3 compounds are reasonable conductors of electricity. They don't compare with really good conductors like copper and silver, but they behave like metals nonetheless. Only experimental evidence concerning the role of oxygen in the material has begun to explain the conductivity in 1-2-3.

Scientists currently explain the electrical conductivity of 1-2-3 in terms of hole conductivity of the oxygen atoms. *Holes* are a name for missing electrons, or electron vacancies like the phantom marbles in our container analogy. Chemical compounds form as a result of individual atoms giving up electrons to other atoms, borrowing electrons from other atoms, or sharing electrons between one another. Such redistribution of electrons results in a lower energy state, so that the compound becomes bound together. In 1-2-3, the oxygen atoms pick up electrons from nearby copper atoms. Despite these additional electrons, the oxygen atom still has room for more in its outer shell; we say that the oxygen still has holes—electron vacancies. Evidence shows that electricity flows through the material by allowing these holes to move from atom to atom in the crystal. In reality, only electrons can actually move around, but there are no itinerant electrons in 1-2-3. The oxygen holes have the freedom to move, so electrons rearrange themselves accordingly. Picture 99 people shifting from seat to seat in a 100-seat theater. The empty seat moves around the room like a hole moves around a crystal. This type of conductivity is not particularly unusual; many semiconductors exhibit hole conductivity.

If we look again at the 1-2-3 unit cell, we notice something else. The all-important copper–oxygen layers are not symmetrically located in the unit cell. In fact, the 1-2-3 material not only looks asymmetric, it behaves asymmetrically. In other words, the compound's properties are different in different directions. In particular, current flows more easily across the copper–oxygen layers than it does along the long dimension of the molecule. Physicists call this kind of behavior *anisotropic* (not the same in all directions) conductivity. The new superconductors appear to be anisotropic in all their superconducting properties. Critical currents, penetration depths, coherence lengths, and other parameters all depend on the orientation of the crystal. The consequences of many of these anisotropic properties remain to be determined, but they already present big stumbling blocks to technologists, because, as we shall discuss later, the ability of the superconductor to carry current strongly depends on its orientation.

The anisotropy of 1-2-3 is not the only problem caused by its complex chemistry and structure. The compound contains chemically reactive components—particularly barium—that strongly interact with other substances. Furthermore, researchers cannot even produce superconducting samples of 1-2-3 without exposing the compound to very high temperatures (700 to 950°C). Without high-temperature processing, the compound does not incorporate enough oxygen into the copper–oxygen layers to produce the right crystal structure. Such temperatures are a problem because they worsen chemical reaction problems with the materials upon which thin films are deposited and, in general, make it difficult to use the new superconductors in conjunction with other materials. These and a host of other materials problems make it difficult to produce high-quality samples of the new superconductors.

Superconductivity above 100 K

Early in 1988, researchers discovered several new oxide superconductors with above 90 K transition temperatures, and these do not even have the same crystal structure as 1-2-3. The newer compounds are still layered copper oxide materials, but they do not contain rare earth elements. In place of yttrium or other rare earths, the compounds contain metals like bismuth or thallium. And unlike either the original Zurich superconductor or 1-2-3, they contain four metallic elements instead of three.

Thus, there exists a whole family of copper-oxide superconductors. In fact, there are at least five different crystal structures that result in high-temperature superconductivity, all composed of layers of copper–oxygen structures sandwiched around layers of other elements. For some reason, having more copper–oxygen layers in the unit cell seems to be conducive to higher superconducting transition temperatures. The superconductor discovered at the University of Arkansas—thallium-barium-calcium-copper-oxide—contains three copper-oxygen layers and superconducts at a temperature thirty degrees higher than 1-2-3.

These newer high-temperature superconductors—both the bismuth-strontium-calcium-copper-oxide discovered in Japan and the thallium compound from Arkansas—create both problems and opportunities for researchers. On the positive side, transition temperatures at 120 K and above mean that these superconductors can actually be used at liquid nitrogen temperature (remember that superconductors are only useful well below their critical temperature). Furthermore, the new materials seem to be more chemically stable and more apt to retain their oxygen than the 1-2-3 compound. On the down side, the newest compounds are even more complicated than 1-2-3, containing five elements instead of four, which increases the challenges faced in producing the material. Also, the thallium compound presents the additional drawback that thallium itself is an exceedingly toxic substance, which greatly complicates its handling.

Most scientists now believe that many more high-temperature superconductors must exist and await discovery. So far, the remarkable string of discoveries that started in Zurich has centered around layered copper-oxide compounds. It may take years to fully explore just these systems, and there may well be other unrelated materials that also exhibit high-temperature superconductivity. Now that the door has been fully opened, there will be ceaseless efforts to find new materials that not only superconduct at higher and higher temperatures, but also ones that can be made efficiently, effectively, and economically. It remains to be seen whether all these new compounds provide new insights about the origins of high-temperature superconductivity, or merely create new problems for theorists. The bismuth and thallium superconductors, for example, do not have the copper–oxygen chains that are present in 1-2-3. Therefore, at least in these newer compounds, the chains do not play a role in high-temperature superconductivity. Clearly, high-temperature superconductivity continues to be a mysterious phenomenon.

The Mystery of High-Temperature Superconductivity

Apart from the practical issue of producing high-quality materials, basic scientific details of the chemical and electrical properties of the new superconductors are not well understood at this point, much less their superconductivity. In particular, why are these new materials superconductors and why are their critical temperatures so high? If ever there was a $64,000 question, this is it. Whoever conclusively answers the question of the origin of superconductivity in 1-2-3 will probably garner whatever fame and fortune that any physicist ever achieves.

As soon as the discovery of high-temperature superconductivity was announced, physicists got busy trying to explain the phenomenon. Every theoretician experienced in superconductivity and plenty of newcomers to the field got into the act. At the American Physical Society meeting in New York in March, 1987, less than 3 weeks after the first publication on 90-degree superconductivity, at least half a dozen different theories were proposed. By June, a conference on novel mechanisms of superconductivity saw 40 presentations attempting a theoretical analysis of the new materials. By early 1988, hundreds of papers had appeared in print dealing with the origins of superconductivity in 1-2-3. Nevertheless, at that point, no consensus had been formed. We don't know exactly why 1-2-3 makes such a good superconductor.

What is the nature of the mystery? Scientists are asking whether the new superconductors are simply particularly good at doing the same thing that the old materials did or whether their superconductivity results from some entirely new mechanism. Put more succinctly, can the BCS theory explain the behavior of high-temperature superconductors?

The evidence so far has been rather mixed. Tests on the new

materials have demonstrated all the hallmark properties of traditional superconductivity: zero resistance, the Meissner effect, flux quantization, and the Josephson effect. Early results have been controversial, but there appears to be an energy gap in the new superconductors as well. Furthermore, flux quantization measurements can identify the charge of the superconducting current carriers. When we make such measurements on ordinary superconductors, we can tell that pairs carry the supercurrents. Significantly, we observe the same thing in 1-2-3. So, from these experiments, we conclude that superconductivity in the new high-temperature materials results from the formation of Cooper pairs, just as it does for other superconductors. Some measurements indicate that the new superconductors do not behave according to the predictions of the BCS theory, but at least the materials appear to superconduct by virtue of the formation of Cooper pair. Nevertheless, we still need to know what mechanism enables the pairs to form.

Perhaps this seems like a minor issue in the greater scheme of things, but the nature of the attractive pairing interaction in a superconductor determines whether the BCS theory can appropriately explain its behavior. As we know, electrons pair up in BCS superconductors because of an electron–phonon exchange process. One electron locally distorts the lattice, a second electron feels the influence of the distortion, and as a result the two electrons are weakly bound together. An important theoretical issue being discussed and debated is whether such an interaction can overcome the effects of electron–electron repulsion and thermal agitation in a material above 90 K. Many theorists think that the answer is no.

If phonons can't cause the electrons to pair up in 1-2-3, then what can? Theorists have found no shortage of alternate candidates. Solid state-physics texts are filled with descriptions of various quantum-mechanical excitations in solids and nearly every one of them has been eagerly embraced by some theorist or another as the new instigator of Cooper pairs. The competition to replace phonons has engaged the participation of scientists in universities and research institutions all over the world. Active support has developed for such mechanisms as excitons, magnons, demons (pronounced "day-mons"), plasmons, superexchange, and resonating valence bonds (RVB). Yet in the midst of all the proponents of these new mechanisms, a vocal minority of traditionalists still clings to phonons as the only mechanism one needs to explain superconductivity.

Perhaps, in the near future, one or more of these new mechanisms will be shown to play the crucial role in high-temperature superconductivity. Then we will have to develop an understanding of the new mechanism akin to our picture of the electron–phonon interaction. Until the favorite emerges, we will continue to let the theorists fight it out on their own. What will settle the fight? More experimental results are needed. We don't know enough about the new superconductors and about their behavior to rule out any of the proposed theories of high-temperature superconductivity. Better experiments on better-quality materials may eventually settle the debate.

Making 1-2-3

The idea of better-quality materials brings us to a topic of greater importance for us than the theoretical understanding of high-temperature superconductivity. We have spent a great deal of time discussing the wide-ranging applications of superconductivity, so we are well aware of the potential that superconductors have for performing useful tasks. As we pointed out earlier, using superconductors requires making useful forms of the materials. More than ever, this represents a tremendous challenge with the new superconductors.

The first samples of the 1-2-3 compound were not wires or thin films, or even hunks of metal. They were little pellets, like aspirin tablets. The first and easiest way to make high-temperature superconductors uses solid-state reaction techniques, or more colloquially, the "shake-and-bake" method. Producing a superconductor in this way is so simple that junior high school students were doing it for science fair projects within a few months of the initial discovery. This was probably the quickest transition from Nobel Prize-winning discovery to schoolhouse science ever to take place.

Making high-temperature superconductor pellets turns out to be more of a home economics assignment than a science project. The starting materials are powdered chemicals containing copper, barium, and your choice of rare earth metal. The original recipe called for yttrium oxide, barium carbonate, and copper oxide. A few dollars' worth of chemicals nets a good-sized pellet of superconductor. If you want to make one, weigh out 1.76 grams of yttrium oxide, 6.17 grams of barium carbonate, and 3.73 grams of copper oxide. The chemicals are not particularly hazardous, but care should be exercised to not ingest them or let them get into the hands of young children. Once the powders are measured out, mix them together very well; a mortar and pestle do the job best. Then bake the mixture in a 950°C oven for 12 hours and allow to slowly cool in the oven. Much lower temperatures simply will not work. While your home oven can't get close to this high a temperature, a pottery kiln can do it easily. The resultant crumbly black substance then has to be squeezed into a solid disk using a die—a metal mold—in a large hydraulic press. Most machine shops have one of these on hand. One capable of 15,000 or 20,000 pounds of pressure will do the trick. The resulting disk gets baked again at 950°C for another few hours, this time in oxygen, if at all possible. Finally, it must again cool slowly in the oven until it can be removed. With this recipe, we can make a 1-inch disk of eighth-inch thickness. With any luck, it will be a 90 K superconductor, made as easy as 1-2-3.

Like any other recipe, this formula for 1-2-3 has its pitfalls. Just like a cake recipe, doubling and redoubling the ingredients often leads to trouble. The temperature of the oven is also critical. The stuff melts just above 1000°C, which ruins the whole process by destroying the crucial crystal structure. In any case, making superconductors this way is so easy that literally thousands of people have done it all over the world. For those of us who don't have access to precision scales, kilns, and hydraulic presses, a number of enterprising concerns have made 1-2-3 pellets commercially available. Six months after the material was discovered, one could buy a 1-inch pellet for about $20, a superconducting bargain.

Once we have one of these pellets, we can do superconducting tricks. Experiments that formerly required liquid helium with its elaborate thermal shielding can be performed on a desktop in a styrofoam cup full of liquid nitrogen. In 1987, we observed superconducting diamagnetism on the evening news as small magnets floated effortlessly above 1-2-3 pellets sitting in liquid nitrogen. How does the floating magnet trick work?

When we place a small magnet on top of a disk of superconductor, the magnetic field tries to penetrate the disk. As we know, the superconductor will exclude the magnetic field by the generation of diamagnetic screening currents at its surface. These currents generate an opposing magnetic field, which in turn repels the little magnet. The superconductor pushes the magnet upward while gravity pulls it back down. The magnet finds a balance between the two forces and stays put, floating in the air.

More elaborate experiments are possible. Several laboratories

A small magnet floats above a superconducting 1-2-3 disk. (Courtesy Lawrence Manning)

have even constructed crude "Meissner motors" that use the repulsion between superconductors and magnets to rotate a small disk on an axle. As time goes on, additional demonstrations will be devised for these easy-to-make, easy-to-handle superconductors. At the very least, there seems to be a rosy future for superconducting toys.

But little ceramic pellets are not the material for future superconducting technology. As we know, most superconducting applications require either wire or thin films; there is little use for pressed-powder pellets beyond scientific measurements and educational demonstrations. The serious development of high-temperature superconductivity will require the serious development of useful forms of the materials.

Taming the New Superconductors

The first question to answer is whether the new superconductors can meet the performance requirements for superconducting applications. Can they carry enough current and stay superconducting in high enough magnetic fields? The first results obtained from ceramic pellet samples were not very encouraging.

Scientists measure a substance's ability to carry current in units of amperes per square centimeter. The copper wires in our walls need to be about $^1/_{16}$ of an inch in diameter to carry household currents. This means that ordinary house wiring carries about 1,000 amperes per square centimeter. Power lines may need to carry as much as 10 times more current per unit area. Conventional wire can seldom handle much more current without elaborate cooling methods. Yet giant power cables are not the most demanding customers for high-current-density materials. The microscopic conductors in integrated circuits often are called upon to carry as much as a million amperes per square centimeter. The actual currents are small inside of microchips, but they may have to flow through conductors that are only ten millionths of an inch wide.

Superconductors have proven to be champion-level performers at carrying currents. Niobium and niobium compounds can carry as much as 10 million amperes per square centimeter. If household wiring was made from superconductors, it would only have to be a hundredth the diameter of standard 14-gauge copper wire, similar to a human hair in thickness. Clearly, high currents present little challenge to the old superconductors.

How do the new superconductors fare at current handling? The early pellets of 1-2-3 could only carry about 1000 amperes per square centimeter before losing their superconductivity. Pessimists came out of the woodwork to announce that the new materials would never be of any use because of their poor current capacity. Not too surprisingly, when better samples of 1-2-3 were made, they were shown to be every bit the equal of the old superconductors in current-handling ability. This should not have been surprising because the pressed ceramic samples really cannot be expected to be able to handle much current.

The microscopic structure of the pellets explains their poor electrical performance. What looks like a solid disk of superconductor is really a bunch of tiny grains that are stuck together. Under a powerful microscope, one immediately sees that the material contains lots of empty space between the grains. Electrical current has to flow from grain to grain in order to travel from one part of the sample to another. The intergrain connections are either formed by extremely small whiskers of material or, worse yet, through layers of insulator.

In fact, current running through a superconducting pellet may have to travel through the equivalent of a treacherous labyrinth of little tunnel junctions. Since such imperfect solids are typical of pressed ceramic materials, we should not be surprised at all that ceramic pellets do not pass large quantities of current. We simply need more refined materials to do that job.

The consequences of anisotropy in current carrying are already well-known. The highest currents are carried by samples of 1-2-3 that have all their crystallites lined up the same way. Scientists call these "oriented" samples. Even better samples consist of just one large crystal. Single crystals and oriented samples make the best conductors because currents never have to flow in the difficult direction along the crystal. Therefore, it makes sense that the random arrangement of the crystallites in a pressed ceramic sample does not make for high current capacity. Only oriented thin films and oriented single crystals of 1-2-3 can carry really large amounts of current. Scientists will need to develop such materials for practical applications.

As far as magnetic fields are concerned, the early results were immediately encouraging. The new high-temperature superconductors are Type II materials, just like the old high-temperature superconductors. And also just like the old materials, the higher temperatures are accompanied by higher critical fields. The highest critical fields estimated for the new superconductors are much greater than can be produced in the laboratory by nondestructive means. At low temperatures, fields as high as two megagauss (4 million times greater than the earth's magnetic field and 2,500 times greater than the critical field of lead) may be necessary to destroy superconductivity in 1-2-3. If 1-2-3 wire can be made into superconducting magnets, the magnets constructed could tear themselves apart from magnetic forces long before the superconductivity of the wire disappeared. So, as far as building magnets is concerned, we just need the right form of the materials for the job.

What are these "right forms" that we keep talking about? Clearly they are wires and thin films of high-temperature superconductors. A major part of the worldwide effort in high-temperature superconductivity consists of trying to produce one or the other of these two forms of the materials. Research in these areas will go on for years to come.

Of the two forms, 1-2-3 wire will be the slowest to get off the starting block. The first attempts consisted of trying to produce coils of ceramic material. Ceramists add organic binders—plastic materials that act as a flexible glue—to the oxide powders to produce a flexible material that they call "greenware." Nonsuperconducting 1-2-3 greenware can be wound into coils, shaped into tapes and rings, or otherwise fashioned into a variety of shapes. The pieces are then heated in an oven—much like the firing of pottery—which bakes off the organic binding material and forms the superconducting compound. The result is a ceramic material much like the pellets but having useful shapes. Such materials suffer from the same problems as the pellets in terms of current-carrying ability, but they are a first step toward wire.

High-current superconducting wire will undoubtedly require large oriented crystals of 1-2-3 in order to overcome the anisotropic conductivity problem. By the end of 1987, Bell Laboratories scientists succeeded in growing long crystals that could handle 10,000 amperes per square centimeter. Such currents are still marginal for most applications, but represent a major step in the right direction. Incorporating large 1-2-3 crystals into wire represents a major technological challenge for the future.

Another wire-making technique involves encapsulating the oxide powders within a copper jacket and then heat-treating them within the copper to form a wire that can be drawn out afterward. Scientists used similar techniques to make early niobium-titanium wires. In the case of 1-2-3 wire, the resultant superconducting core is still a brittle ceramic material, reacts with the copper jacket, and cannot handle much current, but nonetheless represents another approach to the wire problem. Greater success has been achieved by filling silver jackets with already superconducting powder and extruding the material through a narrow opening to fashion wire. Such wire again has poor current-handling ability, but at least the silver does not seem to degrade the superconducting properties of the 1-2-3. Work on many of these wire-forming schemes continues.

Most other serious developmental work on high-temperature superconducting wire has been going on behind closed doors. The payoffs for such a product are potentially gargantuan and it is unlikely that intermediate progress will be given much publicity for fear of revealing proprietary information. When a breakthrough might occur in this area is difficult to assess. It may take months or it may take decades. If the history of superconducting wire provides any clues, researchers will probably need a number of years to develop a successful process for making high-temperature superconducting wire. If motivation and incentives can set the pace, the development of wire will be as fast as humanly possible. Optimists acknowledge the fruits of decades of experience with niobium-titanium and other commercial superconducting wire. Pessimists point to the overwhelming materials problems associated with the new ceramics. The new superconductors are rather crazy materials, by ordinary superconductor standards. Perhaps it will take some crazy ideas to tame them. As usual, only time will tell.

Thin films of high-temperature superconductors, on the other hand, got off to a flying start. The correct chemical formula for the yttrium-based superconducting material was only determined around March 1, 1987. Within 2 weeks, scientists at IBM had made a film of the compound that reached zero resistance at 85 K. Since then, films have been made in numerous laboratories around the world by a wide variety of techniques.

The large number of different film-making approaches has several causes. First, making the films requires elaborate and sophisticated equipment and the quickest route to making films makes use of whatever equipment already exists. Thus laboratories around the world started making films by whatever technique they could immediately employ. The second reason for variety is that no one technique has thus far demonstrated clear superiority over any other. The best way to make 1-2-3 films has yet to be established. Finally, there may be intrinsic advantages to certain techniques for specific applications that don't apply to others.

In light of these facts, we mention just a few of the thin-film processes being developed around the world. Films can be made by two basic processes: starting with the complete compound and somehow depositing onto a substrate or starting with the constituent elements and actually forming the compound on the substrate. The simplest way to make 1-2-3 thin films is to start with a pellet of the material and somehow atomize it. One can do it with a laser, with electron-beam evaporation, with sputter deposition, or even with a vapor spray nozzle. In each case, the liberated 1-2-3 material then lands on a substrate material and forms a thin film coating. Researchers produced vapor sprayed films of 1-2-3 quite early on. The technique produces rather rough coatings of the material, but they do indeed superconduct.

The next easiest technique turns out to be the laser. The method, called *laser ablation*, produces films by blasting off tiny amounts of superconductor from the pellet with an energetic laser beam. Fairly smooth and high-quality films can be made by this method. Sputtering from a pressed powder ceramic sample has also proven to be successful. Because of uneven sputtering of the compound, researchers typically have to sputter specially prepared starting materials that compensate for this effect by containing extra amounts of harder-to-sputter elements. Identifying the correct targets requires much time and effort, but Japanese physicists have made great progress with this thin-film technique. Electron-beam evaporation of pressed pellets of 1-2-3, on the other hand, seems doomed to failure. The intense beam of electrons heats a tiny spot on the pellet and causes it to melt and then vaporize. This violent process invariably tears the compound apart so that it typically does not redeposit in the proper elemental proportions.

A different approach to electron-beam evaporation has nevertheless proven to be very successful in making 1-2-3 films. In fact, electron-beam evaporation was the first method used to make films by IBM. Its method was to use three beams at once and separately evaporate yttrium, barium, and copper. If the amounts of each metal are correct and a little oxygen is added, one can make 1-2-3 by this technique. Some of the best films continue to be made by the three-electron-gun evaporation technique and by a related method called *molecular beam epitaxy*. Some laboratories have achieved their best results by evaporating barium fluoride rather than barium metal in their systems.

Researchers have also applied the three-metal method of making films to sputtering. Three sputter guns simultaneously deposit yttrium, barium, and copper in the presence of oxygen to produce the desired compound. As in the three-metal evaporation technique, the real challenge in multigun sputtering is the careful control of the amounts of each of the metals being deposited. This method also produces high-quality films in a growing number of research labs.

Many other film-making processes are being developed as well. In a number of labs, researchers are trying to implement complex chemical processes including techniques called *metal-organic chemical vapor deposition* (MOCVD) and sol-gel casting. We won't detail these or many others here. Suffice it to say that experts from many areas are trying their hands at making high-temperature superconducting films.

The best films of 1-2-3 made during the first year since the discovery do not compare in quality with films of niobium, niobium nitride, and other older superconductors. The films do not have completely uniform composition, are not microscopically smooth, and react with the substrates upon which they are deposited. More seriously, 1-2-3 films also tend to have thick layers of nonsuperconducting compounds on their surfaces. Such unwanted material is caused by some of the elements in the film migrating to the top and forming undesirable compounds. Problems of this sort prevent the immediate development of a high-temperature superconducting electronics technology. Before such a technology comes about, scientists will have to find a way around the material-specific problems of high-temperature superconductors.

On the positive side, after only 1 year, tremendous progress had been made on the development of high-temperature superconducting materials. Researchers had produced thin films with 90 K superconductivity and high current-carrying ability. More and more was learned about the nature of the materials and the means necessary for producing them. And all the while, eager scientists and technologists awaited better and better samples for their research. The discovery of other, higher temperature superconductors early in 1988 sparked additional excitement in an already exuberant field. Researchers found themselves with several different materials to choose from for the development of high-temperature superconducting technology. Within weeks of the later discoveries, scientists had already made thin films of the bismuth-based superconductors. Apart from the race to discover new materials and develop the ones that have been discovered, a competition begam among the new superconductors to determine which one(s) will become technologically important. The practical issues of making high-quality superconducting materials will decide which superconductor wins out in the end.

Next, we look at the scope of the worldwide effort in high-temperature superconductivity and we assess the state of the high-temperature superconductor revolution.

CHAPTER 15

A Super Opportunity

We have traced the recent events that resulted in the discovery of superconductivity above 77 K, the temperature of liquid nitrogen, and we have examined the nature of these new compounds and the major challenges they present. Now we want to assess the impact that these new superconductors are likely to have in the future and try to understand why their discovery has triggered something of a revolution in the scientific world. In other words, why is high-temperature superconductivity such big news?

Unquestionably the greatest single consequence of the discovery of high-temperature superconductivity has been an enormous increase in the public's interest in superconductivity. New books are being written, courses being taught, conventions being held, and generally the whole subject of superconductivity has risen from almost total obscurity to high visibility.

At this point, we look into three issues related to the new-found celebrity of superconductivity. We show why superconductivity above the temperature of liquid nitrogen makes such a difference, we assess the impact of high-temperature superconductors on superconducting applications, and we discuss what actions governments, private companies, and universities around the world are taking in response to the discovery of high-temperature superconductivity.

Why Liquid Nitrogen?

According to the popular press, liquid-nitrogen-cooled superconductors have great potential because liquid nitrogen is so cheap.

Everyone points out that while liquid helium costs as much as decent wine, liquid nitrogen prices out like cheap beer. This may be true, but it doesn't reveal some of the most significant advantages of 77 K superconductivity.

Perhaps the most important advantage involves the latent heat of vaporization of liquid nitrogen. We talked about latent heat before; it is a quantity of energy that must be added to a substance to complete a phase transition at the critical temperature. In this case, the latent heat corresponds to the amount of energy that must be added to liquid nitrogen at 77 K in order to make it boil away. Liquid nitrogen has the technological advantage that this energy is 60 times greater than the amount needed to vaporize liquid helium; boiling off nitrogen is much harder to do. What do we gain from this difference?

A simple demonstration makes the point. If we fill up an ordinary coffee thermos with liquid nitrogen in the evening, the next morning there will still be liquid nitrogen in it. The large latent heat of nitrogen means that the stuff does not boil away very easily. Liquid helium, on the other hand, boils so readily that we can hardly pour it fast enough even to form a level in any open container. Without extra layers of insulation or even a liquid nitrogen jacket, liquid helium boils away almost immediately.

A number of applications require superconducting equipment to endure long periods of unattended operation in out-of-the-way or even totally inaccessible locations. Obvious examples are buried SQUID sensors, satellite-borne electronics, and energy storage coils in isolated locations. Everything else being equal, liquid-nitrogen-cooled superconductors would require refills of refrigerating liquid 60 times less often than helium-cooled superconductors. A SQUID magnetometer that currently requires a monthly refill of helium would run unattended for 5 years with liquid nitrogen. Thus high-temperature superconductivity promises to deliver a huge gain in adaptability to long-term operation.

A second benefit derived from 77 K operation comes from the reduced power requirements for cooling. Refrigeration uses up energy. A sizable chunk of our monthly home power bill goes to running our refrigerator. Many potential users of superconductivity have to address the energy costs associated with cooling. Anyone who wants to operate electronics on a satellite has a keen interest in this problem; there are no extension cords out in orbit. Circuits that operate at 77 K require far less energy than circuits running at liquid-helium temperatures. Engineers measure cooling capacity in power units—watts—just like they measure heating capacity. In this case, the cooling power measures how fast energy can be taken out of a material in order to lower its temperature. It takes 1,000 watts to run a 4 K refrigerator that can cool at a rate of 1 watt; it takes 40 watts to run a 77 K refrigerator that cools at the same rate. Therefore, cooling a material with liquid nitrogen is 25 times more energy efficient than cooling it with liquid helium. In power-poor environments like aboard a spacecraft, this is a major consideration. In fact, the liquid-nitrogen temperature range also opens up some attractive alternatives for cooling. A number of efficient self-contained refrigeration machines are already available for the 60 to 80 K temperature range. With these compact refrigerators, cryogenic liquids like liquid nitrogen might not even be necessary to provide cooling for superconducting circuitry. Satellites may even be able to maintain such temperatures nearly for free. By radiating heat out into space, orbiting satellites can passively reach temperatures as low as 100 K, leaving very little work for an active refrigerator.

A third important consideration results from the tremendous simplification in special handling required for liquid-nitrogen-cooled systems as compared with liquid-helium-cooled systems. Once again, a thermos full of liquid nitrogen can vividly demonstrate this fact. Liquid helium can only be stored in expensive, elaborate containers and requires specialized equipment just to transfer the stuff from one storage vessel to another. Liquid nitrogen, on the other hand, can be poured like coffee. Using nitrogen drastically reduces the amount of gear needed to stay cold.

So from a design perspective, 77 K superconductivity offers an enormous reduction in the complications inherent in operating in a low-temperature environment. Whether we are discussing large superconducting magnets in levitated trains or magnetic resonance imaging machines or superconducting integrated circuits on board satellites, a technology based upon liquid nitrogen cooling represents a giant step toward practicality.

These issues represent the underlying justification for the tremendous excitement generated by the discovery of high-temperature superconductivity. Nonetheless, the magnitude of the worldwide reaction to the discovery has far outstripped people's understanding of its technological benefits. We now examine why this should be.

Impact of the Discovery

Two aspects of the discovery of high-temperature superconductivity have greatly heightened its impact: it was essentially unexpected, and yet its potential was already well understood. Both of these factors have led to tremendous enthusiasm in the scientific community that has rapidly spread to the business community, to government agencies, and to the general public.

The scientific community did not expect high-temperature superconductivity to happen, yet in some sense, it seemed to be waiting for it anyway. Virtually every university, government, and industrial laboratory that was involved with superconductivity immediately recognized the significance of the discovery and starting working with the new superconductors. In many cases, existing research projects were abruptly shelved to make room for the new activity. No one in the field needed to be convinced that liquid-nitrogen-cooled superconductors were important and deserving of study. Researchers everywhere just jumped right into the fray.

The almost-overnight mobilization of the scientific community was an impressive phenomenon. In December, 1986, perhaps a couple of dozen researchers worked on the copper-oxide superconductors; by March, 1987, at least 3,000 scientists were involved. By 1988, as many as 30,000 scientists were pursuing high-temperature superconductivity research. Several motivating factors could be seen at work.

On the level of pure science, nothing has more appeal than a new phenomenon that needs explaining. Early in 1987, everything about the high-temperature superconductors still needed to be learned; there was room for researchers in physics, chemistry, crystallography, material science, and several other disciplines. The pickings were incredibly ripe for new discoveries, publications, and professional advancement. Alex Mueller and Georg Bednorz were awarded the 1987 Nobel Prize in physics for their discovery of the first of the new high-temperature superconductors. The onslaught of superconductivity research in 1987 eventually yielded some 3,000 published articles appearing in hundreds of journals around the world.

The sheer volume of research going on in superconductivity was enough to grab the public's interest. People wanted to know what all the excitement was about. As a result, superconductivity became a very visible scientific field and its principles and properties became the subject matter of magazine and newspaper articles and television reports. Ironically, apart from reporting on the remarkable discoveries taking place, most of the news coverage centered around long-established information about superconductivity. Yet to the general public, all of it was news.

This sudden "coming out" of superconductivity represents one of the most interesting aspects of the high-temperature superconductivity revolution. We would almost think that an entirely new phenomenon had been discovered when, in fact, the reality was that a 76-year-old phenomenon had just become much more practical. All of a sudden, ideas that had been developed over decades became exciting new innovations.

People immediately wanted to know what they could do with the new superconductors. The answers are obvious. The applications they read about in the magazines are many of the same ones we have explored throughout this book. We can easily envision extrapolating the low-temperature superconducting technology to the new materials. In most cases, higher temperature operation can enhance superconductivity's advantages. Nevertheless, we should not lose sight of an area of even greater promise for the new materials. These compounds have unusual properties and physical structures that present tremendous practical challenges to superconducting technologists. However, these same unique features may very well lead to entirely new applications of superconductivity, ones based on devices that have not even been conceived of yet. The long and diverse list of applications of superconductivity has taken many years to amass; we have only just begun to think about the potential of the new superconductors.

As far as the impact of the new materials on existing technology is concerned, the prospects are very exciting. Naturally, all hopes rest on our ability to produce useful forms of the superconductors—high quality thin films and practical wires and cables. But assuming that these materials challenges can be met—a task that may well occupy the better part of a decade—a great potential for application awaits.

Putting the New Superconductors to Work

In the area of large-scale applications of superconductivity, economics rather than potential performance has always been the driving force. Designers of particle accelerators and medical imaging machines are already willing to pay the price for superconductivity; as we have seen, power and transportation engineers have not been so willing. Would these applications become economically viable with the advent of liquid-nitrogen-cooled superconductors? Very likely, they would. Economy of scale serves as a guiding principle. A general rule states that higher temperature superconductors can achieve commercial viability in smaller scale systems.

As we have already seen, the utilities industry promises to be an area of high impact for superconductivity. While practical nitrogen-cooled superconductors are not likely to stimulate a wholesale restructuring of the electrical power system, many applications will become very attractive. Since superconducting power cable can carry far more power than its conventional counterpart, we may use it in situations where the right-of-way for additional power lines cannot be obtained. Superconductors offer the option of getting more out of the existing rights of way. Another way in which nitrogen-cooled superconductors can make immediate economic sense is the opportunity to locate controversial power plants—for example, nuclear power plants—very far from populated areas and transport the electricity over large distances. Resistanceless superconducting cable makes this idea feasible. Energy storage systems offer a third immediate application for nitrogen-cooled superconductors. In late 1987, the United States Department of Defense contracted to develop a test model of an SMES (Superconducting Magnetic Energy Storage) system of the type we discussed earlier. The teams selected to build the system will initially work with commercially available low-temperature superconductors, but they will continue to monitor the development of the new materials for future energy-storage applications. Since energy-storage coils are limited to a fixed location and can be designed without major constraints upon size, weight, and configuration, SMES systems figure to be one of the first places where large-scale applications of high-temperature superconductivity make a significant impact.

Motors and generators constructed from liquid-nitrogen-cooled superconductors are likely to be smaller, more efficient, and more reliable than the conventional machines they would replace. Such superconducting rotating machinery could eventually replace most conventional large-scale electrical machinery. Small-scale machinery would probably still not gain enough in performance to warrant the use of superconductivity. Such applications may have to await the discovery of even higher temperature superconductivity.

Superconductivity offers several specific advantages for transportation systems. Magnetic levitation provides a means of achieving high-speed ground transportation. A system based on nitrogen-cooled superconductors might well be economically viable; a system based on higher temperature superconductors, were they to be found, would be almost indispensable. The weight and configuration advantage of superconducting motors for shipboard propulsion are also well known. The economic equation comes out about the same as for ground transport. At 77 K, it would probably be worth doing; at much higher temperatures, it would be inevitable.

Apart from opening up new markets, nitrogen-cooled superconductors would greatly increase the market penetration of the existing applications. The use of high-temperature superconductors would decrease the cost and complexity of medical imaging equipment and nitrogen cooling itself would make such systems far more user-friendly. Similarly, the rather substantial refrigeration costs incurred by large-scale particle accelerators would be substantially reduced by liquid-nitrogen cooling. The existing market for large superconducting machinery will undoubtedly greatly expand with the commercialization of the new superconductors.

Superconducting electronics also stands to gain from the development of the new materials. The most impressive performance may come from large-scale electronic systems like the Josephson junction computer. But projects of such complexity may take some time to develop. Smaller scale electronic applications could well be the first commercial applications of the new superconductors. Predicated on the development of high-quality thin-film materials and Josephson junctions, circuitry based on nitrogen-cooled superconductors would have many ready-made applications in high-speed and high-frequency electronics.

SQUID magnetometers for medical applications would be far easier to operate and could even be brought closer to the patient on account of the reduced bulk of the thermal shielding for the sensors. Other SQUID applications requiring long-term deployment would clearly benefit from higher temperature operation, since they could function unattended for years if necessary. If sufficiently sensitive and noise-free SQUIDs can be made from the new superconductors, look for a major expansion of the commercial use of SQUID electronics.

Superconducting microwave and infrared detectors, mixers, and other microwave components are already finding a home in a growing number of communications, surveillance, and scientific applications. All these endeavors would benefit from the new higher temperature materials. In fact, the use of high-temperature superconductors will increase the range of useful operating frequencies of these devices to unprecedented values. Analog- and digital-signal processing electronics stand to make the same gains in the transition to high-temperature superconductors. If the technological accomplishments in conventional superconductors can be repeated with the new materials, new standards of high-speed electronics performance will be set. There may be tradeoffs to be made in terms of ultimate low-noise and low-power performance by high-temperature superconductive electronics, since higher noise levels inevitably accompany higher temperatures and higher signal levels accompany the large energy gaps of high-temperature superconductors. Nevertheless, a high-temperature superconductive electronics technology would unquestionably have a great deal to offer. It is probably safe to assume that the high-performance, high-frequency electronics of the future will probably make use of superconductivity in one way or another.

Another area of tremendous promise that has come to the forefront because of the new superconductors is the idea of hybrid superconductor–semiconductor technology. The marriage of superconductivity with conventional semiconducting electronics was never considered to be practical because of the extreme low temperatures required for superconductivity. Most semiconductor devices simply cease to function at liquid-helium temperature. But why should we want to use superconductors and semiconductors together? The answer is symbiosis; each technology has something to offer that the other cannot.

One attractive option is to use superconductors to interconnect the semiconducting devices in conventional circuitry. When the number of devices crammed into a tiny circuit becomes sufficiently high and the size of the conductors becomes sufficiently small, superconducting interconnections become very advantageous. A semiconductor-based computer that uses superconducting wiring could be miniaturized to an extent that ordinary copper wiring would not permit.

Other ideas that combine superconductors and semiconductors divide up the tasks in a complex electronic system according to the talents of the participating technologies. In an advanced computer, for example, Josephson junction circuitry has been shown to be superior for the implementation of a high-speed central processor. Memory circuits based on Josephson junction circuitry, on the other hand, have not progressed as far. One can easily imagine building a supercomputer that combines both Josephson circuitry and advanced transistor circuitry in order to exploit each technology to the fullest extent.

As we have already seen, superconductivity currently accounts for about $400 million in annual sales, although most superconductor products are part of larger systems that cost considerably more. The commercialization of high-temperature superconductors will undoubtedly bring about substantial growth in the superconductivity industry. Applications that previously were not economically viable will become attractive, and established applications will experience far greater market penetration. If even higher temperature superconductors were discovered and commercialized, the technological consequences would be enormous. If room-temperature superconductors came on the scene, they would rapidly become all-pervasive in a diversity of applications areas.

Meeting the Challenge

Here, at the beginning of the era of high-temperature superconductivity, scientists and technologists are faced with a series of significant challenges to be met before any of the promise of the new materials can be realized. Without functional wires or Josephson junction devices, the new superconductors cannot be put to work. The rapid but somewhat directionless mobilization of the worldwide scientific community must now be channeled into activities that will lead to the rapid development of the new technology. The industrialized nations of the world are now actively turning to this task.

The worldwide effort to promote high-temperature superconductivity centers around four broadly defined objectives. These include the development and improvement of our understanding of the new materials; the development of practical forms of the new superconductors; the fabrication of prototype devices, circuits, and machinery based on the new materials; and the continuing search for higher temperature superconductors.

Universities, government labs, and industrial organizations have all actively embraced the task of expanding our understanding of the new superconductors. The work seeks to answer two basic questions: what are the superconducting and normal-state properties of the new materials and what basic mechanisms are responsible for the high-temperature superconductivity? During the first year since the discovery of the new compounds, the tremendous and essentially spontaneous redirection of effort on the part of the scientific community has not been accompanied by a tremendous redirection of financial resources. Late 1987 saw a great deal of talk about additional money for basic research in the United States, but the traditional sources of funds—government agencies—were only beginning to create new programs.

The discovery of high-temperature superconductivity has been hailed as a triumph for "small science." In an era filled with multi-million-dollar gigantic scientific machines staffed by dozens or even hundreds of researchers, small teams of individuals working with table-top equipment have done revolutionary work. High-temperature superconductivity was discovered by individual effort and ingenuity rather than by the commitment of mammoth resources. Development of our basic understanding of superconducting phenomena may well continue to take place in the same domain.

But even if individual effort continues to pave the way for new breakthroughs in superconductivity, it will no longer be the isolated efforts of a handful of researchers. Thousands of scientists now study the properties of high-temperature superconductors around the world. In the United States, scarcely a university exists without some sort of research effort in the new field. All the major government research laboratories have large and active programs for studying high-temperature superconductivity.

In Japan, where applied-superconductivity research already enjoyed heavy governmental support, there has been a strong push to increase government spending on basic research. Somewhat stung by a much-discussed reputation for mainly developing the discoveries of other countries, the Japanese government now pushes a "Be Creative" campaign to stimulate basic research and development work. Already the acknowledged leader in developing technology, Japan is determined to establish itself as a world leader in basic scientific research as well.

In Europe, there has also been a spontaneous shift toward high-temperature superconductivity research. University and industrial laboratories have re-programmed research funds into the new efforts and various collaborative efforts are at least in the planning stages. In the USSR, where basic superconductivity research has been a major activity for 50 years, high-temperature superconductivity has also attracted a great deal of attention. While some Soviet work has been reported at international conferences during 1987, the full scope of the research efforts on the new materials has not been made public.

The Chinese have played an active role in the discovery and development of the new materials since the very beginning. Superconductivity research has become a major scientific activity in China and Chinese scientists are increasingly becoming frequent participants in the international scientific scene. All around the world, scientists continue to try to understand the properties of and the underlying mechanisms behind the new high-temperature superconductors.

While scientists pursue their inquiries, and government agencies around the world debate the merits of increased support of basic research in superconductivity, research on superconducting technology continues to experience enormous growth. In the United States, the charge has been led by government research laboratories and major industrial research labs.

As we have seen, the original discovery of high-temperature superconductivity came from a small research laboratory operated by IBM in Zurich. The subsequent breakthroughs occurred at university laboratories in Houston and Alabama. Following the initial discoveries, the first large-scale efforts to produce technologically important high-temperature superconducting materials have taken place at the major research laboratories of IBM, ATT Bell Laboratories, and Bell Communications Research, as well as at government facilities such as the Naval Research Laboratory, Argonne National Laboratory, and Los Alamos National Laboratory.

By the time a year had gone by, many other industrial and government laboratories had become involved in the new superconducting technology development effort. Major aerospace firms, communications companies, chemical manufacturers, and electronics manufacturers all initiated programs aimed at the technological development of high-temperature superconductors. In the past, superconductivity research in the United States has mostly been funded by a variety of government agencies, particularly research arms of the Department of Defense, NASA, and by the National Science Foundation. In the wake of the high-temperature revolution, American corporations have begun to pour their own money into the effort. By 1988, worldwide funding for superconductivity research had reached the level of $350 million a year, an enormous increase over previous levels.

After 1 year, researchers had quite a ways to go before they could begin to build real devices, machines, and circuits from the new superconductors. Several laboratories were able to demonstrate rudimentary SQUIDs from both pressed pellets and thin films of 1-2-3. The devices did not demonstrate impressive performance, but scientists were encouraged nonetheless. If poor SQUIDs can be built, then good SQUIDs are at least possible. As of early 1988, no real Josephson junctions had been made from the new superconductors. Producing a sandwich structure containing two 1-2-3 thin films presents many technological challenges. As thin-film quality improves, many laboratories continue to try to develop high-temperature Josephson junctions. Each small step toward working high-temperature superconductor devices will continue to attract a great deal of attention over the next few years.

Searching for New Superconductors

We have touched upon the first three objectives of the research effort on high-temperature superconductivity. We have talked about basic scientific research, about applied technology research, and about the work on devices. The fourth research objective is the development of still newer superconductors.

By the time of Mueller and Bednorz' discovery, hunting for high-temperature superconductors had come to be an activity pursued only by a handful of diehards. Very little research funding was earmarked for the purpose, and few expectations were held for any results. Since the discoveries of 1987, the search for higher temperature superconductivity has taken on a whole new respectability and even some glamour. The frenzy experienced in early 1987 gradually died away, but most large research institutions continued to pursue new materials at least at some level.

The breaking of the liquid-nitrogen-temperature superconductivity barrier has shattered psychological barriers as well. The second wave of superconductor discoveries in early 1988 yielded new families of above-liquid-nitrogen temperature superconductors, still based on copper oxide but no longer containing rare earth elements. These newer compounds superconduct at higher temperatures than 1-2-3 (over 120 K), and open up whole new areas for research. Future research will determine whether the newer compounds or the rare earth compounds prove to be the best materials for applications.

Most scientists now fully expect that even higher temperature superconductors—perhaps even room-temperature superconductors—will eventually be discovered. Before Mueller, Bednorz, and Chu made their discoveries, only the most wildly optimistic researchers made such claims. The thallium-based copper-oxide superconductor broke the 125 K barrier in March, 1988, and scientists fully expect the 150 K barrier to fall with the same class of compounds. In the 2 years beginning in 1986, superconductivity climbed from very low temperatures to halfway to room temperature. The race to push temperatures higher will continue to be a part of the greatly enlarged field of superconductivity research.

We should say a little about room-temperature superconductivity, a phrase that pops up in conversation with increasing frequency. In the practical sense, a room-temperature superconductor would not be a substance whose resistance disappears at 68°F. We have seen that superconductors only become useful when they are operated well below their critical temperature—one-half to two-thirds of that temperature provides a rule of thumb. At higher temperatures, superconductors can carry very little current and cannot withstand strong magnetic fields. Therefore, for the technologist, a room-temperature superconductor would be a substance whose resistance disappears somewhere above 450 K (270°F). Such a material could actually be used at room temperature.

Where Do We Go from Here?

An interesting offshoot of the major intensification of superconductivity research has been the growth of interest in conventional (liquid-helium-cooled) superconducting technology. The benefits of superconductivity are becoming well known to a much broader audience than ever before. Potential users who never before considered superconducting technology are now aware of the advantages of superconducting devices, circuits, and machines—even when they operate only at liquid-helium temperature.

Two factors accentuate the new interest in the old superconductors. First of all, people are starting to realize that it may be some time before the new high-temperature superconductors are ready for real applications. They may not be willing to wait several years and they may want to test out their ideas at lower temperatures in the meantime. At the very least, low-temperature superconductivity is a vital proving ground for all superconducting applications. There exists far greater motivation for demonstrating functioning superconducting applications at low temperature; scientists now have good reason to believe that the same tasks could eventually be made to work at liquid-nitrogen temperature. In the past, this was only an unjustifiable pipe dream.

The second factor encouraging the use of low-temperature superconductors is the enabling nature of the discovery. The discovery was a discovery in science, not in technology. New applications did not suddenly come into existence, only a new viability of the existing applications. Therefore, the increased visibility of superconductivity in general has created a renewed interest in superconducting applications that is quite independent of the applicability of the new materials. We can expect the development of applications based on conventional superconducting materials to proceed at a much higher rate than ever before.

Apart from this coat-tail effect for low-temperature superconductivity, high-profile programs in high-temperature superconductivity have sprung up in a great number of industrial laboratories. High-temperature superconductive electronics technology is the target for many of these programs. Therefore, many parallel efforts are now underway to develop thin films, Josephson junctions, and other superconducting circuitry based on the new materials.

The United States government, although slow to release new funds, has taken an active role in promoting rapid progress in superconductivity research. At a conference sponsored by the federal government in June, 1987, President Reagan called for a concerted team effort by American government and industry to establish leadership in superconducting technology. Congressional legislation, agency policy decisions, and funding programs all seek increased cooperation on the part of research institutions. As a result, universities, corporations, and government laboratories are pursuing a variety of cooperative activities in support of superconductivity research. The Japanese efforts in superconductivity continue to serve as a model for these programs.

There has been no shortage of entrepreneurial activities in superconductivity either. A sizable number of new companies has sprung up almost overnight to take advantage of various aspects of the boom in superconductivity. These businesses address a variety of interests, ranging from information services to the development of specific technologies.

By early 1988, over a dozen new publications had come on the scene to serve the growing superconductivity community. These include a number of weekly, bi-monthly, and monthly newsletters and magazines (sporting names like *Superconductivity News*, *Superconductor Week*, *Supercurrents*, and *High Tc Update*), several new scientific journals entirely devoted to superconductivity, and some high-priced research studies on various business opportunities related to superconductivity. Also pertaining to the expanding superconductivity information industry, several superconductivity trade associations have been formed and have been instrumental in the organization of nontechnical conventions and other activities devoted to superconductivity.

On a more technological level, a number of companies have been established for the manufacture and sale of high-temperature superconductor products. In early 1988, these primarily consisted of powdered materials and pressed pellets, presumably catering to the needs of the educational and research communities. Of interest in these endeavors is the fact that high-volume production techniques for the new superconductors have been developed in a rather short time period. This fact bodes well for whatever commercial applications researchers develop for the new materials. We can only speculate about the future plans and activities of companies that are currently producing superconducting powders and pellets.

On a more substantial level, at least by monetary standards, are a number of venture-capitalized startup companies in superconductivity. Millions of dollars have been committed to a number of new firms that have announced their intentions to develop products based on high-temperature superconductivity. In most cases, these companies have affiliated themselves with respected scientists from the world of superconductivity and have launched initial programs of research aimed at identifying the most promising product areas. The long-term plans of these companies have probably not yet been formulated and it will undoubtedly be at least one or more years before any of these firms begins to pursue their goals in earnest. In the meantime, there appears to be no shortage of investors eager to take part in the superconductor revolution and we should expect to see the establishment of many more new superconductivity companies.

Whereas the question on the public's mind and on the minds of funding agencies is: how long will it take to develop the new superconductors into useful devices, circuits, and machinery, the question on the minds of most scientists working on the new superconductors is: how much time will the world give us to do the job? The eventual success or failure of high-temperature superconductivity may well depend upon how far apart the answers to these two questions turn out to be.

For many years, the superconducting research community has worked on science and technology on an annual budget of tens of millions of dollars. Current plans call for hundreds of millions of dollars to be spent on the intensified effort in superconductivity research. Such a increase in funding will be accompanied by at least as large an increase in expectations. Whether researchers can deliver on schedule remains to be seen. In the past, developing new superconductors for specific applications has been a slow process, one fraught with disappointments and dead ends. The sheer volume of work now being done on the new materials may well shorten the gestation period for the new technology, but the work will still take some time. The unprecedented complexity of the new materials coupled with the unprecedented payoffs envisioned for a successful development effort will create a high-pressure environment in the research community for years to come.

Crystal-ball gazers of all persuasions are now looking intently at the future of superconductivity. What will the next year bring? What will the next decade bring? The only certainty seems to be that there will be tremendous activity in superconductivity research. Thousands of scientists are hard at work investigating the new superconductors, developing them for applications, searching for higher temperature materials, and putting low-temperature superconductors to work in a variety of applications. Some scientists are even taking the opportunity to think of innovative new applications that take advantage of the novel properties of the new materials. Whatever happens with high-temperature superconductivity, it will take place under intense public scrutiny. Superconductivity has lost its cloak of obscurity and has become a very visible area of science. The coming years will determine whether it will live up to its new-found fame.

CHAPTER 16

Connecting the Future

As recently as 1986, most scientists doubted that they would ever find a room-temperature superconductor. But now such a discovery seems more and more likely. We can't be completely sure someone will in fact develop a room temperature superconductor, particularly one that engineers can successfully shape into wires, circuit components, and other useful things. But suppose someone finds one, and suppose that in time scientists figure out how to synthesize it in useful forms and build devices out of it—say, in less than 25 years from now in 2011, in time for the celebration of the 100-year anniversary of Kamerlingh Onnes' discovery of superconductivity. What technological marvels could we expect to see?

Assuming that the room-temperature superconductor could be made economically in abundant quantities, there would be no drawback to using it in practically every industry and technology of modern society, since its use would not involve cooling costs. The benefits would range from minor improvements in existing technology to revolutionary upheavals in the way we live our lives.

A world with room-temperature superconductivity would unquestionably be a cleaner world and a quieter world. Superconducting electronic motors would replace many noisy, polluting engines. Compact superconducting cable would replace unsightly power lines. Advanced transportation systems would lessen our demands on the automobile. And energy savings from many sources would add up to a reduced dependence on conventional power plants. The effects of room-temperature superconductivity would be felt throughout society.

Room-temperature superconductivity would revolutionize the electrical power industry. A practical room-temperature superconductor would virtually make obsolete the use of ordinary copper for carrying electricity. There would simply be no reason to use it any longer. All the dreams of miles of superconducting wire would finally come true. Since there would be no need to cool the wire, cables could be very thin, so existing rights-of-way for power lines could carry many times larger amounts of power than they do now. And of course, since we would not lose power to resistance, extremely long runs of transmission line would at last be practical. We could put just a few fossil-fuel burning power plants, or nuclear power plants, or other sorts of power plants that might be unsafe, dirty, or otherwise undesirable for populated areas, on distant and isolated sites, and transport the electricity efficiently over thousands of miles, if necessary, to where people live. Thus we could keep the inhabited areas clean and safe.

Superconducting magnetic energy storage would become commonplace, and we could couple it to alternative energy sources to make them really useful. Power sources whose operation cannot be synchronized with demand—such as windmills and solar collectors—would deliver their energy to superconducting storage coils. These coils would save the energy until it is needed. Of course, storage coils would also handle the excess output of conventional generators—which would themselves be improved by superconducting wire. In this way, the electrical power grid could bank its energy reserves for times of peak demand.

The end of copper's dominance in electrical wiring would carry over to the home as well. Household wiring would scarcely occupy any space at all, since superconducting wiring with a very tiny diameter could do the job. All the large, heavy motors around the house would be exchanged for light and compact superconducting replacements. Thus there would be a new generation of smaller refrigerators, washing machines, dryers, and air conditioners. And lightweight, powerful superconducting motors would show up in small appliances such as mixers, blenders, hair dryers, and fans as well. There would probably be little reason to remove all the old copper wiring to put in superconductors, but nothing new would be made with copper any longer.

New and newly improved conveyances would dominate the world of transportation. Trains, highways, private automobiles, and ships would change dramatically. The magnetically levitated trains that offer advantages of tremendous speed and quiet operation would become eminently practical without the need for cooling the superconducting magnets. Advanced systems requiring superconducting magnets in both the train and the track bed would deserve serious consideration. In the most ambitious design, travelers could cross the American continent in a few hours aboard a supersonic maglev train encased in an evacuated above-ground tube. They could cross the country by train as fast as, or faster than, we now do by airplane. Shorter maglev train routes, for example between major cities or from cities to resort areas, would pop up everywhere, because they would be cheap to run, and they would be comfortable, quiet, and fast.

On a smaller scale, individual transportation might bear little resemblance to today's automobile. Futuristic cars could benefit from superconductivity in a number of ways. The same magnetic levitation principles used for the railroads could be put to work on superconducting superhighways. Cars could enter special on-ramps, pick up their wheels, and be carried away by linear induction motors in the maglev highway. Accidents would be a thing of the past, as the cars would move at high speed to the synchronized pulse of the magnetic control system. Drivers would travel as fast as 150 mph on a true superhighway and never need to worry about collisions.

Leaving the magnetic highway, cars could take advantage of other advanced systems. Ultrasensitive, superconductive, high-frequency electronics could turn the lowly car into a sophisticated computerized transportation system with advanced navigation capabilities. Each car would have a powerful Josephson-junction computer in the dashboard. Drivers would simply tell the computer where they want to go. On the roof of the car, a dinner plate-sized satellite dish (small because a weak signal suffices for the ultrasensitive superconducting receiver) would link the computer with an orbiting navigational satellite—also packed with superconductive electronics—that would precisely pinpoint the car's current location. By continuously updating this information, coupled with extensive geographical data in the computer's memory, the computer would direct the car to its destination. Thanks to the high-frequency, high-sensitivity operation of superconductive electronics, no driver need ever get lost, even in an unfamiliar city.

As the car makes its way through a maze of city streets, the same sort of superconducting magnets used for highway levitation would now serve as effective safety devices, as they generate strong magnetic fields that repel other similarly equipped vehicles, creating a magnetic cushion between adjacent cars. Even on a more mundane level, the car would be filled with superconductors. All the electrical motors in the car—from windshield-wiper motors to fan motors to power-window motors—would be replaced by lighter, more powerful superconducting motors. Even the generator would be wound with superconducting wire. Whether cars would run on electricity or on gasoline, they would feature a number of superconducting systems.

Out at sea, ships propelled by superconducting motors would dominate the world's commercial and military fleets. Propulsion systems would weigh less and use space more effectively than on present-day ships, and the ships would be far more maneuverable. The entire drive system from generator to power cable to motor could be made with room-temperature superconductors, with nary a cooling system in sight. And, of course, the superconductor-based navigational systems that go into cars would certainly be a fixture aboard these future ships.

SQUID sensors would become ubiquitous in many areas of technology. Advanced medical diagnostic instruments would become compact, portable devices. Susceptometers and magnetometers would be built into hand-held units that could rapidly locate dysfunctions in many body systems. Without cumbersome thermal insulation and cooling systems, SQUID sensors would be as simple to use as a stethoscope. A compact, dedicated Josephson computer would reduce the data from the SQUID circuits into a form that doctors could readily use.

A combination of powerful, miniature superconducting motors and sophisticated Josephson computer circuitry could lead to advanced prosthetic hands and limbs whose functions would rival the "bionic" body parts of popular fiction. Powerful motors would no longer have to be bulky, inconvenient machines; they could be incorporated directly into the prostheses. Thus, an amputee could receive a new hand with strong, grasping fingers that he or she could control with great accuracy.

The same miniature motors would revolutionize precision tools of many kinds. Professionals and hobbyists could use small, hand-held electric drills, saws, screwdrivers, and lathes that are lightweight and maneuverable and more powerful than today's bulkier tools.

SQUIDS and other superconducting detection devices would become the sensory organs of advanced new industrial robotic devices. With superconducting sensors to see the world, superconducting motors to operate their limbs, and a superconducting computer to control their action, such future machines would begin to resemble the robots we have always imagined. Unlike today's massive, awkward industrial robots, these would be compact and facile.

Freed from the encumbrances of refrigeration, superconductive electronics would become a powerful presence in communications, information processing, consumer electronics, and a host of other areas. Superconducting circuits would be smaller, faster, quieter, and more power efficient and would operate at higher frequencies than any conventional technology.

Room-temperature superconductive electronics would open up frequencies one hundred times larger than now practical for commercial exploitation. With such frequencies we would have an enormous assortment of new satellite channels available for commercial broadcasting, telephone systems, and navigational and other specialized communication systems. Also, sensitive superconductive electronics would dramatically reduce the size of the equipment needed for satellite communications, so more electronics could be packed into an average-sized satellite, and we could replace those large rooftop receiving dishes with tiny ones.

Josephson electronics would dominate the world of mainframe computers and might easily take over the small computer market as well. Powerful portable computers that run on a small battery would be possible with low-power superconducting circuitry. Superconducting magnetic memory could revolutionize the whole technology of information storage. Digital data could be stored in units of single flux quanta contained in tiny superconducting loops, which would enable engineers to build incredibly dense memory circuits that can be erased and rewritten. This would allow us to store a lot more information in less space than we can with any other magnetic recording system. Moreover, superconducting shielding would protect such data from external magnetic influences. Such conventional information storage systems as magnetic recording tape and computer disks could become obsolete.

Superconductivity would find its way into high-performance consumer electronics. Digital audio equipment with 10 times today's sampling frequencies would ensure flawless reproduction of the most demanding musical signals. Loudspeakers with superconducting voice coil wiring would be virtually indestructible no matter how loud the music. Finicky hi-fi buffs would even use superconducting speaker cables to ensure that the purest signals reach their speakers. Finally, superconducting infrared sensors would produce remote control devices for home electronics that would work from almost any position and distance.

We could even imagine that inventors would tap superconductors for new recreational activities. Children of the next century might well grow up playing with superconducting toys—levitating spaceships, magnetic table games, and intricate motorized toys of many kinds. Perhaps room-temperature superconductors would inspire a new kind of hockey in which players skate upon a superconducting rink and attempt to hit a floating magnetic puck into the enemy goal: levitated super-hockey. Given new freedom, inventors would have a field day.

All these innovations are almost obvious ways to use room-temperature superconductors. If we gave such a material to a roomful of engineers and industrial designers, and told them to come up with something new, there is no telling what they might invent. Room-temperature superconductivity would undoubtedly trigger a revolution of scientific imagination.

The vision of the future that we have portrayed is merely science fiction at this point. However, it differs from ordinary science fiction in an important way. We have not invoked notions of time travel, antigravity, or greater-than-light speed; we have not pictured a world with alternate laws of nature. Instead, we have speculated about the impact of the expanded role of an established phenomenon. Superconductivity at room temperature would not put us on another world, but it would make this world a different place in which to live.

We may or may not find superconductors that work at room temperature, and if we do find them, we may or may not be able to fashion them into useful things. The quest for such materials will certainly go on in the closing years of this century. With new superconductors being discovered all the time, we have good reason to be optimistic that the search will someday succeed.

In the meantime, researchers continue to work with existing superconducting materials to devise and perfect a rich variety of useful machines and devices. In the years to come, with or without room-temperature materials, superconductivity will play a growing role in how we obtain and consume energy, how we get from place to place, how we communicate and process information, how we care for the sick, and how we alter and protect the environment. In short, superconductivity will have an increasing impact on our overall quality of life. Even at this moment, we can think of superconductivity as a technological link connecting us with the future.

Glossary

1-2-3—A class of superconducting ceramics, such as yttrium-barium-copper-oxide, with critical temperatures in excess of liquid nitrogen's boiling point. The name "1-2-3" refers to the atomic ratio of number atoms in the chemical formula, $Y_1Ba_2Cu_3O_7$.

Alternating current, ac—Electrical current that periodically reverses its direction of flow.

Analog electronics—Electronics that handles signals in the form of voltages that directly replicate the original information.

BCS Theory—A theoretical explanation of superconductivity first proposed by Bardeen, Cooper, and Schrieffer. BCS theory predicts how the forces between electrons and surrounding atoms can lead to the pairing of electrons, resulting in a zero-resistance state of matter.

Coherence length—The fundamental size scale of superconductivity. In a superconductor, paired electrons are separated by, on average, the coherence length.

Cooper pair—Two electrons bound together by interaction with the atomic lattice. Cooper pairs are responsible for superconductivity.

Coulomb force—The electric force that repels like charges and attracts opposite charges.

Critical current—The maximum current a material can carry in its superconducting state.

Critical field—The maximum magnetic field in which a material can remain superconducting.

Critical temperature—*See* Transition temperature.

Current—The flow of electrons. Engineers measure current in amperes (amps), one of which is equivalent to the flow of roughly 10^{19} electrons per second.

Diamagnetism—Tendency of a material to exclude magnetic field lines. Superconductors are perfect diamagnets, excluding all magnetic fields from their interior.

Digital electronics—Electronics that handles signals in the form of binary numbers (bits) that form a snapshot of the original information.

Direct current, dc—Electrical current that flows in a single direction through a circuit.

Electric field—A spatially extended region surrounding a charged particle that exerts a force upon any other charged particle.

Electron-beam evaporation—A thin-film deposition technique that uses a focused beam of electrons to heat and vaporize a target material.

Electron–phonon interaction—The interplay between electrons and the atomic lattice, responsible for both ordinary resistance and superconductivity.

Energy gap—The forbidden energy zone for individual electrons in a superconductor. Additional energy, at least equal to the gap energy, must be supplied to break Cooper pairs into independent electrons.

Energy level—In quantum mechanics, an allowed value of energy that a particle can have within the confines of a physical system.

Fermi level; Fermi energy—The highest occupied electronic energy level in a material.

Flux quantum; Fluxon—The smallest nonzero amount of magnetic flux that can exist in a superconducting ring. Superconducting rings can hold only an integer number of fluxons. A fluxon is an extremely small amount of flux, equivalent to about a millionth of the earth's magnetic field covering a square centimeter.

Flux, magnetic—A measure of the strength of a magnetic field through a specific area.

Inertia—The tendency of a stationary object to remain stationary and for a moving body to keep moving along the same direction.

Ion—An atom with either missing or extra electrons. Atoms become ions when they form a crystal lattice.

Isotope effect—The influence of atomic mass on the superconducting transition temperature.

Josephson effect—Any one of a number of consequences of Brian Josephson's theory for the behavior of superconductivity in tunnel junctions. Cooper pairs can carry current across a tunnel junction with no applied voltage (dc effect). Applying a dc voltage to a tunnel junction produces radio waves (ac effect).

Josephson junction—A superconducting tunnel junction consisting of two superconductors separated by a very thin insulating barrier. Both individual and paired electrons travel from one superconductor to the other, crossing the insulator in a process called tunneling.

Joule heating—The generation of heat associated with the flow of electrons through a conductor.

Kelvin—A temperature scale used by scientists in which absolute zero is 0 K and other temperatures are defined as 273° above Celsius temperatures.

Kinetic Energy—Energy of motion.

Lattice, crystalline lattice—The orderly arrangement of atoms in a solid.

Lenz's Law—A changing magnetic field will induce a current in a conductor whose flow generates an additional magnetic field that resists the changing field.

Liquid helium—A 4 K fluid that can be used to cool superconducting materials and circuits.

Liquid nitrogen—A low-temperature fluid that provides cooling at 77 K.

London penetration depth—The short distance that magnetic fields can penetrate into a superconductor.

Maglev; Magnetic levitation—A lifting process made practical by powerful, lossless superconducting magnets.

Magnetic field—The region of magnetic force surrounding a magnet or current-carrying wire.

Magnetometer—An instrument for measuring magnetic field strengths.

Maxwell's Theory—A set of equations describing the relationships between electric and magnetic forces and charges.

Meissner effect—The expulsion of magnetic field lines from the body of a superconductor when it becomes superconducting, also known as perfect diamagnetism.

Mixed state—Magnetic behavior of Type II superconductors in which applied magnetic fields produce a microscopic array of superconducting and normal regions.

Momentum—Mass in motion; the product of mass and velocity.

Monopole—A theorized magnetic particle having only one magnetic pole.

MRI; magnetic resonance imaging—A medical diagnostic technique, similar to X rays, but using radio emissions caused by shifting magnetic fields.

Niobium—A hard, grayish metallic element used on its own and in compounds for superconducting applications.

Normal—Nonsuperconducting (as applied to metals).

Ohm—The unit of electrical resistance. An ohm corresponds to 1 volt per ampere.

Ohm's Law—The current that flows through most materials is proportional to the voltage applied across the material.

Pauli exclusion principle—For the class of particles called fermions, which includes electrons, no two particles can exist in the same quantum state.

Perfect conductor—A conductor with zero resistance. Perfect conductors can pass currents with no voltage present.

Persistent current—The steady flow of electrical current that can be maintained indefinitely in a loop of superconducting wire without an applied voltage.

Phase—The stage in the time development of a wave or wavefunction.

Phase transition—A change from one thermodynamic state to another, such as the transition from liquid to solid or superconducting to normal.

Phonons—The natural vibrations of a solid's crystal lattice.

Photolithography—A technique of patterning thin films based upon light-sensitive, removable coatings.

Picosecond—One trillionth of a second.

Potential energy—Stored energy available for use. Energy stored in a compressed spring or gravitational energy stored in an uphill reservoir of water are two examples of potential energy.

Quantum mechanics—The modern description of the behavior of matter on the atomic level.

Quantum state—The set of quantum mechanical properties that specifies a particular particle, wave, or system; typically includes energy, momentum, spin, etc.

Quench—Sudden loss of superconductivity in cable or magnet due to thermal "runaway."

Rare earth—Any of over a dozen silvery-white, soft metallic elements including lanthanum, yttrium, and a number of others.

Resistance—The tendency of materials to obstruct the flow of current; quantitatively, the ratio of the voltage across a conductor to the current flowing through it.

Room temperature superconductor—In practical terms, a material that has useful superconducting properties at 300 K; would require a critical temperature of at least 450 K.

Semiconductors—A class of solids with electrical properties in between conductors and insulators, used for solid state electronic devices.

SMES; superconducting magnetic energy storage—A system to store electrical energy using current in superconducting coils.

Sputtering—A thin-film deposition technique using projected ions to knock atoms away from a target of source material.

SQUID; superconducting quantum interference device—SQUIDs are electronic circuits consisting of one or more weak links, often Josephson junctions, interrupting an otherwise superconducting loop. SQUIDs are extremely responsive to small magnetic signals.

Superconductivity—An electronic state of matter characterized by zero resistance, perfect diamagnetism, and long-range quantum mechanical order.

Supercurrent—A resistanceless current carried by Cooper pairs.

Thermodynamics—The science that relates states of matter to heat and other forms of energy.

Transition temperature; T_c—The temperature at which a material enters the superconducting state.

Transistor—The key circuit element in semiconductor technology that can be used for amplifiers or switches.

Tunnel junction—A structure consisting of two metals separated by a microscopically thin insulating barrier that can pass electrical current only by a strictly quantum-mechanical process (see Josephson junction).

Two-fluid model—A theoretical description of superconducting phenomena based on the coexistence of superconducting electrons and normal electrons within a superconductor.

Type I—Category of superconductors characterized by perfect diamagnetism until a critical magnetic field is reached, at which point the material turns normal.

Type II—Category of superconductors in which magnetic fields can penetrate the material in an orderly array of normal regions containing magnetic vortices.

Unit cell—The basic building block of a crystalline solid.

Valence electrons—The electrons in an atom that determine its chemical properties and participate in the formation of molecules and compounds.

Voltage—Electromotive force; the driving energy that acts to push electrons through a material.

Wavefunction—The mathematical description of a quantum mechanical entity—such as a single electron, or all the Cooper pairs in a superconductor—that predicts its behavior.

Weak link—Part of a superconducting circuit intentionally made to have weaker superconducting properties than the rest of the circuit—often a Josephson junction.

Further Reading

General Superconductivity

Pierre DeGennes, *Superconductivity in Metals and Alloys* (New York: W. A. Benjamin, 1966).

Alistair Rose-Innes and E. H. Rhoderick, *Introduction to Superconductivity* (London: Pergamon Press, 1978).

Michael Tinkham, *Introduction to Superconductivity* (New York: McGraw-Hill, 1975).

Large-Scale Applications

Simon Foner and Brian Schwartz, eds., *Superconducting Machines and Devices: Large Systems Applications* (New York: Plenum Press, 1974).

Martin Wilson, *Superconducting Magnets* (New York: Oxford University Press, 1983).

Superconductive Electronics

Antonio Barone and Gianfranco Paterno, *Physics and Applications of the Josephson Effect* (New York: John Wiley & Sons, 1982).

Theodore Van Duzer and Charles Turner, *Principles of Superconductive Devices and Circuits* (New York: Elsevier, 1981).

Solid State Physics

Neil Ashcroft and N. David Mermin, *Solid-State Physics* (New York: Holt, Rinehart and Winston, 1976).

Charles Kittel, *Introduction to Solid-State Physics* (New York: John Wiley & Sons, 1976).

Historical Background

Robert Hazen, *The Breakthrough: The Race for the Superconductor* (New York: Summit Books, 1988).
Fritz London, *Superfluids, Volume I: Macroscopic Theory of Superconductivity* (New York: Dover, 1961).
Kurt Mendelssohn, *Quest for Absolute Zero* (New York: McGraw-Hill, 1966).

Index

[Page numbers in *italic* indicate illustrations]